THE THAMES AND HUDSON MANUALS
GENERAL EDITOR: W. S. TAYLOR

Architectural Ceramics

David Hamilton

The Thames and Hudson
Manual of Architectural
Ceramics

with 146 illustrations in colour and black and white

Thames and Hudson

Printed in Great Britain by
Cox & Wyman Ltd,
London, Fakenham and Reading

Contents

Acknowledgments

I would like to record my thanks to the many individuals
and firms who have provided assistance, information and
illustrations, in particular Mr David Malkin, Mr Ron Leak
and Mr George Price of H. & R. Johnsons Ltd, Mr Peter
Ainsworth of Hathernware, Mr Ward and Mr Coucil of
Shaws of Darwin, Mr R. Griffin of Allied Insulators Ltd,
Mr Hewson of Redbank Manufacturing Co. Ltd, Mr P.
Smith of John Caddick & Sons Ltd, Miss Loveridge of
Doulton & Co. Ltd, the staffs of the Gladstone Pottery
Museum, the Department of the Environment and the
Greater London Council Photo Archives, Dalit Daruvar
and Dr S. B. Hamilton OBE for his advice on Coade stone.

Ruth Duckworth, Peter Edwards, Francis Hewlett and
John Mason were all kind enough to provide me with
photographs of their work, for which I am most grateful.
I would like also to thank Sandra Taylor for allowing me to
photograph some of the tiles in her collection, Peter Bridge-
water for providing drawings from some very slight
sketches, Frank Thurston for translating my negatives into
prints and Irene Owen who typed the text from tapes and
almost indecipherable notes. I am also indebted to the staff
and students of the School of Ceramics at the Royal College
of Art for all their assistance in putting me into touch with
manufacturers as well as clarifying various technical matters.

This book is dedicated to Jan, Sarah, Jonathan and Olivia
for all their help and support at a difficult time in their lives.

Chiswick 1977

Introduction

The term 'architectural ceramics' has become so widely used to describe a variety of otherwise unclassified ceramic objects that it has almost come to mean all things to all men. For the purposes of this book it applies only to those clay items which constitute part of a building or are of such a scale that they may be regarded as existing within an architectural environment and making a substantial contribution to it.

Therefore the techniques described relate to the methods of making bricks, tiles, faience and terracotta, and large-scale sculpture, as well as roofing tiles, chimney pots and other decorative features. I do not distinguish between factory and studio production as many factories concerned with this type of work would regard themselves, quite rightly, as being involved with craftsmanship and, except in the case of automated production, the making processes do not vary so much whether the site is a factory or a studio, or whether the product is made by many people or just one or two.

As to who will use this book, time alone will tell: it is certainly aimed at expanding the horizons of students studying ceramics. I hope that it will illustrate that the activity is much broader than the mainstream work of pottery and, in more recent times, small sculpture. There have been many occasions in the past (and some of these not so long ago) when 'decorative ceramics' meant more than just the trading of fancies, when it was integrated into an architectural activity concerned with the visual quality of architecture as well as the economy of building. Some of the heroic work of the past may serve as a beacon to those who recognize that decoration is not synonymous with depravity or decadence but is a striving for the humanity in architecture which should be the motivating force in any building design. Already some techniques are almost forgotten but it may be possible to revive these if the climate becomes more amenable, although if such a revival is to come about it should not be an insipid revamping of earlier styles but rather a vigorous and vital feature of a new decorative architecture.

1 'Large Hand' by Francis Hewlett. The hand, which is 150 cm (5 ft) high, was built up by coiling.

1   History of architectural ceramics

The first architectural ceramics were almost certainly produced in the Near East in the fifth millennium BC when the brickmakers in that area decided to strengthen the unfired clay bricks, bonded with straw, by firing them, probably in an open fire. They soon recognized the advantages of applying potters' techniques, and decorative walls, modelled and glazed, have been found and dated as having been made around 1400 BC. Within 200 years the Egyptians had developed the manufacture of tiles to the point where they were acceptable to the Pharaoh for the decoration of his palace. By 580 BC the Babylonian Empire had developed an architectural style involving the depiction of mythical animals, modelled in relief on large brick walls and, like the Egyptian tiles, revealing a variety of colours. But whereas the Egyptian tiles used the inlay technique and clays of different colours to depict the image, the Babylonian walls included glazes of different colours. In China by the third century BC tomb figures and tiles were in common use.

Brick- and tile-making spread throughout Europe under the Greek and Roman Empires. The latter in particular established brick- and tile-making skills wherever their armies penetrated. But their architectural purpose remained functional rather than decorative. It was some time after the decline of the Roman Empire that there was any attempt to imitate the Roman mosaics by the use of decorated floor tiles.

Various decorative techniques, including *cuerda seca*, were developed in the Near East and spread through Africa with the Moorish invasion which reached as far as Spain and France. With the establishment of Islam in the tenth century the discouragement of natural forms in any artifact produced a flourishing geometric style of decoration, utilizing the knowledge of mathematics which had all but disappeared in Europe during the Dark Ages but was preserved and developed in the Near East.

Relief tiles and bricks were produced in northern Europe with the development of a permanent architecture and, in particular, within fireplaces when suitable natural materials were not available. By the fourteenth century the Moors had carried their skills in architectural ceramics as far as Spain, for instance in the Alhambra. At this time encaustic tiles, rich in heraldic devices, were in common use throughout northern Europe, particularly for the floors of churches and abbeys.

2 Dutch brick fireplace back. Brouwershuis, Antwerp.

3 Thirteenth-century encaustic tiles from the Chapter House, Westminster Abbey. One fish head is a nineteenth-century copy.

4 Panel of dark-blue tiles portraying the entombment of Jesus. This is part of the tiled wall of the Martyrs' Chapel, Fronteira, Spain. Workshop of Oliveira Bernades, *c.* 1717.

The fifteenth century in Florence saw the development of terracotta, majolica or faience; some of the richest examples come from the della Robbia workshops, which produced large, modelled, polychromatic, semi-structural panels in perfect harmony with the painting and architecture of the period.

Persia in the fifteenth century produced lustre tiles which have seldom, if ever, been surpassed. Under Shah Abbas the imperial mosques and palaces were decorated with painted and mosaic tiles and glazed bricks which by this time included a floriated design.

As the social structure in Europe developed into a more permanent fabric so the architecture involved more permanent materials which came to be regarded as a sign of status. Wealthy landowning families required houses which entailed a little fortification but more and more decoration. Decorative terracotta façades were a feature of fifteenth-century Italian architecture which was carried through Europe along with other cultural and political influences. During the reign of the Tudors in England this use of decorative terracotta and brickwork was much in evidence. In the seventeenth century, even glazed house signs were a feature of Italian domestic architecture. By the eighteenth century floors, walls and ceilings of certain rooms in Dutch houses can be found clad entirely in tiles; the decorative technique which became increasingly popular was the painting of a flat tile with under-glaze or in-glaze colours. In church architecture throughout Holland, Portugal and Spain patterned tiles enriching an otherwise barren surface developed into grand *trompe l'oeil* illustrations of stucco architectural detail, or took the place of paintings, with a subject matter suitable for the environment. These were often carried out in monochrome, usually blue and white.

With industrialization and the growth of towns during the late eighteenth and early nineteenth centuries, brick-making became an increasingly important industry. The demand for readily available, durable and cheap building materials became an economic necessity, and a determining factor in the growth of cities and towns throughout the United Kingdom. By the mid-nineteenth century the increasing enthusiasm for medieval culture, partly as a reaction against the classical style of building, produced a Gothic Revival architecture which created a demand for terracotta façades, encaustic and tessellated pavements, as well as elaborate floor tiles. This style of building persisted alongside the 'classical' style of architecture, but the supporters of each influence fought constantly for ascendance.

With the growth of cities in America, similar architectural styles were adopted and terracotta façades, as on the Woolworth building in New York, were made possible by the influx of Italian craftsmen skilled in terracotta techniques.

5 Entrance to the Natural History Museum, South Kensington, London. Polychrome faience tiles.

6 Office interior by the American Art Tile Company (early twentieth century): relief tiles on the floor, glazed wall tiles and fireplace.

7 Tile with engraved outline, transfer-printed and then filled in with colour by hand.

Such was the boom in the building industry that only factory manufacture could meet the enormous demand, and in the tile industry methods of tile-making and decoration were the subject of much research and mechanization. The development of printing for transfer and direct printing on to tiles, together with mechanical plastic and dust pressing of the tiles themselves, were major factors in the widespread use of tiles within domestic buildings.

The reaction against industrialization, in the later nineteenth century, known as the Craft Revival, includes the work of William de Morgan who, obsessed with the re-invention of Islamic glazes and lustres, devoted considerable energy to the manufacture of painted lustre tiles for interior surfaces of houses and even ships.

By the 1920s and 1930s the most popular and characteristic ceramic façades are found in cinemas, particularly of the Odeon chain, and in large public buildings.

At present, tile production is centred in the United Kingdom upon the do-it-yourself market. The installation of thick or large tiles requires the services of a tile-setter,

8 Melias Building, Liverpool. Polychrome faience tiles.

but the sale of tiles which can be fixed by any handyman with the latest adhesives was identified as a potentially large market, and manufacturers determined the design of the tiles to facilitate this type of setting. Only in the manufacture of tiles for swimming pools and prefabricated architectural panels have some of the richer tile surfaces been preserved. Extruded tiles, in particular, have been much used in recent years, originally set on to on-site cast concrete surfaces but more recently set into pre-cast concrete panels fabricated under factory conditions.

In countries such as Italy with a long tradition of tiled surfaces within domestic environments and which have methods of house construction which lend themselves to tile cladding, nothing like the variety that might be hoped for has been achieved. Manufacture in Italy includes large quantities of unglazed and undecorated tiles which are sold to many decorating workshops, both large and small, each with a particular style or ceramic quality. These are marketed by encouraging the public to replace tiles as fashions change. The misconception that ceramic floors are hostile and cold in domestic environments has not prevented their use on a massive scale in Italy and other countries. This cannot be attributed to climatic conditions, as in some areas of Italy, at least, the climate can be as severe as any found in northern Europe or northern America. Italian manufacturers find a ready export market for their products, although these are considered to be expensive compared with some other tiles.

The number of brick and tile manufacturers in the United Kingdom has declined considerably during the last seventy years, but the level of production has not similarly declined. Successful manufacturers have bought out their competitors and continued to manufacture within the factories which have come under their control, at least for short periods. This increasing concentration of products under the control of two or three major manufacturers has produced a market situation where there is little variety of design, each successful product being quickly copied by competitors in the same market. Invention and creativity are concentrated on the means of production rather than the product itself. Visiting any major manufacturer of ceramic tiles, one cannot fail to be impressed by the ingenuity shown in devising equipment and machinery for handling the tiles throughout the complex process of manufacture. In recent years, however, markets have been subjected to increasing competition from manufacturers in other countries offering either a different style or a richer ceramic quality. As industrially produced objects become increasingly expensive owing to vast capital investment and increasing rates of pay for employees, so the hand-made, or studio-based, product can be expected to become increasingly competitive, and a ready market may well be found for a product with a different appearance or ceramic quality.

Opposite:
The Ishtar Gate from Babylon (now in the Staatliche Museen, Berlin). Built by Nebuchadnezzar (605–562 BC). Polychrome, modelled brick, 40 ft high (12.2 metres).

College of the Mother of the Shah, Isfahan, 1706–14.
Detail of the glazed brick and faience dome.

Glazed earthenware figure of a Judge of Hell. Chinese,
16th century.

Other types of architectural ceramics may well be dependent on buoyant economic conditions. It is regrettable that in any architectural scheme the first feature to be discarded, in order to save money on the project, is usually the creative, decorative element which is within the scope of one artist/craftsman, or a small studio employing four or five people. There are signs, however, of a more enlightened attitude among architects and clients, where such items, within an architectural scheme, are no longer considered as decorative extras but may be the only part of the design which the users of the building find significant and bearable. In several countries, including some states within the USA, there are schemes such as the one called 'one per cent for art' which demands that one per cent of the over-all contract price be devoted to permanent works of art. In this situation ceramics is an ideal medium providing for enormous variety of surfaces and durability. There are encouraging indications that among architects and planners these qualities have aroused an increasing interest in the use of ceramics in architectural environments.

9 Dresden Bank, Chicago. Hand-modelled stoneware tiles by Ruth Duckworth, 1977.

2 Clay

Clays are formed by the decomposition of felspathic rock under the weathering action of wind and rain or erosion by rivers. Clays which decompose upon the site of the original rock are known as primary clays and are characterized by relative purity with a lack of plasticity: they may not hold their shape during making and firing, but tend to crack. They are relatively refractory and fire white or light cream in colour. Clays which have been transported from the site of decomposition and deposited on the beds of rivers or lakes are known as secondary clays. These are far more plastic but relatively impure, because during their passage they will have taken up impurities present in other rocks, which have similarly decomposed. Nevertheless, both these types of clay are given the standard chemical formula of $Al_2O_3SiO_2H_2O$. They are broadly categorized as aluminosilicates, and from the chemical formula it will be apparent that they include chemically combined water. The impurities associated with secondary clays tend to reduce the refractoriness of the clay, and the fired colour may be whitish, through buff to dark brown depending in part on the proportion and type of metal oxide incorporated into the clay.

Microscopic examination shows that the clay particles take on a plate-like form, and the arrangement of these platelets will have a substantial effect upon the physical characteristics of the clay during making. Secondary clays, by the process of sedimentation, tend to have a large proportion of their particles lying parallel to each other, while primary clays display a much more random arrangement. During deposition of secondary clays there is a tendency for the larger particles to fall away first, and for the lighter, smaller platelets to be carried farther downstream. Thus secondary clays may display a more limited variation in particle size than primary clays.

WATER IN CLAY

Two types of water are to be found in workable deposits of clay, chemically combined and physically combined. The physically combined water, which is driven off during drying and firing, surrounds the platelets of clay and enables them to move independently of each other; thus the clay may be formed and hold its shape during drying. This lubricating water may be present in sufficient quantities to allow the platelets to flow, and in this condition the clay is known as a slip or slurry. In every process of ceramic manufacture some physically combined water will be present in the clay. It is this water, acting as a lubricating agent, which allows the clay to be shaped and to hold its form. Without it the clay would be a dry dust which would simply fall apart once the forming forces had been withdrawn.

DRYING CLAY

Even when the clay appears totally dry or de-watered there will be some water remaining within the interior. The drying action is dependent upon the evacuation of the spaces (capillaries) between the clay particles and, under normal conditions, this capillary drying breaks down before the clay is completely dry. It is therefore necessary to raise the temperature of the clay very slowly so that the water may be driven from the interior of the clay before it assumes a vapour state, at 100° C (212° F) and beyond. If all the physically combined water is not driven off, it will become steam, and generate such enormous pressures that the clay will be shattered by the increase in volume of the water as it is converted to steam. Once the temperature of the clay reaches 300° C (570° F), the chemically combined water starts to be driven off, and this process is normally complete by the time the temperature reaches 500° C (930° F).

The increase in temperature during this part of the firing must be very slow indeed in order to avoid the total destruction of the clay form.

FIRING CLAY

As the temperature of the clay continues to increase it undergoes certain changes. Some of the silica which is present in the clay is in the form of free silica, i.e. it is not combined chemically with the alumina. This free silica undergoes an increase in volume as it passes 573° C. This change is known as the alpha/beta conversion, the silica being defined as 'alpha' silica below 573° C and 'beta' silica beyond it. As the temperature of the clay moves to 1100° C the beta quartz is converted very slowly to beta cristobalite and expands in volume once more. Not all the beta quartz converts immediately, and the proportion of beta cristobalite present in the clay is dependent upon the final temperature achieved and the time taken to reach it. The conversion of quartz to cristobalite is irreversible, but the conversion of alpha quartz to beta quartz is reversible upon cooling. Once the firing is complete the cristobalite ceases to be converted and at 573° C the remaining beta quartz reverts to alpha quartz and contracts. At 220° C the beta cristobalite inverts to alpha cristobalite and contracts.

These changes in volume of silica present in the clay can cause cracking, or 'dunting', of the ceramic form if the critical temperatures are passed too suddenly. This is particularly important with large ceramic objects as it is difficult to maintain an absolutely even temperature throughout the volume of the kiln. Should the free silica present in the clay convert only in one part of the ceramic form, then the change in volume in that area may be sufficient to cause a crack between it and the rest of the clay object which is at a different temperature.

Throughout the heating-up process other changes occur within the clay form. Any vegetable matter within the clay will be burned and give off gases, so it is important that the kiln is well ventilated in the early stages of firing. Above 800° C the alkalis present in the clay act upon the silica and alumina to fuse the clay progressively as the temperature rises. The particles of clay thus start to bind themselves together, in a process known as sintering. As the temperature continues to increase, the alkalis act further on the silica and alumina to form a glassy matrix which serves to bond the undissolved particles together.

The higher the temperature to which the clay is fired the stronger it will be upon cooling. If the clay reaches a state where sufficient glass is formed within it to prevent water impregnation when it is cold, the clay is said to be 'vitrified'. Naturally as the materials melt within the body, they flow into the open spaces between the clay particles resulting in a shrinkage of the clay object. Also, the higher the temperature to which the clay is fired the more it will shrink.

Clays which include a smaller proportion of alkalis or other minerals to act as fluxing agents upon the alumina and silica are said to be 'refractory', but any clay will melt, and deform to become ultimately a glassy puddle, if it is subjected to sufficient heat. The most common fluxing agents occurring naturally within clays are sodium, potassium and calcium.

If several clays are combined, the resulting mixture is known as a 'body'. Such combinations are normally effected where a single clay will not provide all the characteristics required in the making or appearance of the finished object. One notable inclusion in the case of large-scale ceramics is the addition of grog (fired clay crushed and ground to a convenient particle size) in order to reduce the shrinkage of the clay during firing and drying, and to increase the particle size and the size of the capillaries within the clay body, thus making easier the evacuation of the water from the interior of the clay object.

Clays are found in combination with many other materials, and vary considerably in composition. Some of these combinations render the clay unsuitable for any process of manufacture except brickmaking. For the purposes of this book it is only necessary to distinguish between two types of clay. The first is clay which is suitable for large-scale production of bricks; the second is those clays which, in combination, lend themselves to the production of tiles, architectural faience and terracotta, porcelain and stoneware forms. All the latter are normally produced by combining clays from several deposits, together with other minerals, in such a way that they produce the desired working and fired characteristics.

Whether the clay being used is a compounded body or a naturally occurring clay it should be fired at least once during the firing cycles to a temperature which will provide good mechanical strength. All clays have a maturing temperature at which the clay develops the maximum strength without being deformed.

PARTICLE ORIENTATION

While the clay is being prepared for use it is most likely that it will at some time pass through a pugmill. If the clay is wedged or otherwise broken up to dissociate it from this pugging process, no evidence remains of the spiral orientation created by the augers or blades which mix the clay and drive it through the mill. In some manufacturing processes the clay is not changed substantially after being pugged, and this causes problems in the most diverse methods of forming the clay – in insulator factories where the clay comes from the pugmill to be turned, and in twin-tile production and roofing and ridge tile extrusion, where the clay is extruded to form it into the desired shape. The particles tend to orientate themselves around the axis of the augers, thus producing spiral laminations. During drying this spiral may show signs of unwinding so that the extrusion will twist and warp out of shape. If the clay could be extruded as a homogeneous mass completely free of laminations or regular orientations many of these problems would be eradicated, but no process has yet been devised which will improve the clay as a pugmill does without the pugmill's disadvantages.

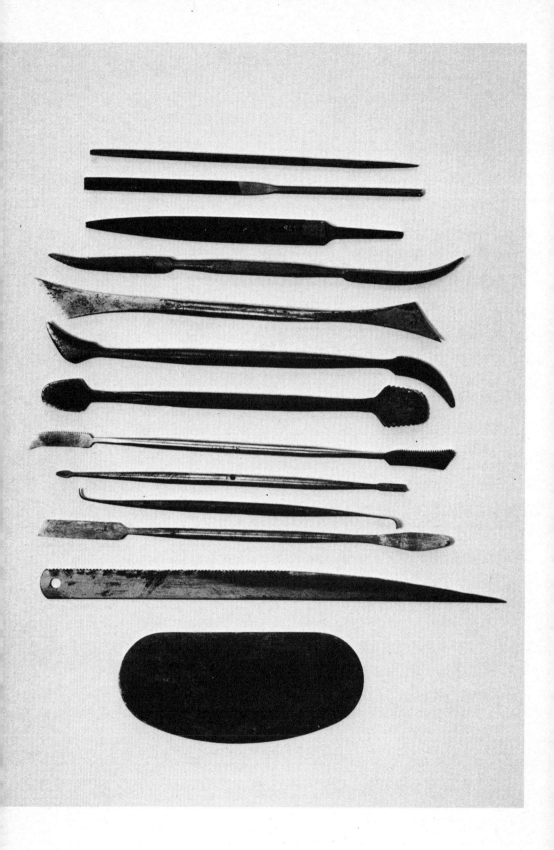

3   Basic techniques of forming clay

Before proceeding to make ceramics of an architectural nature it is necessary to have some experience of the basic hand making techniques of fabricating clay forms, as it is extensions of these techniques which are used in architectural ceramics. There are four methods of making ceramic objects which should be experienced on a relatively small scale if you have little or no experience in making ceramics.

COILING

For this process you need a firm table, a small board covered with a piece of hessian or polythene, and a spatula-type modelling tool. Take a lump of clay weighing four or five pounds, of the type which will produce the characteristics which you require and will withstand the firing temperatures of the glazes you wish to use. This clay is rolled out upon the table and if the surface is such that the clay starts to stick to it, then it will be necessary to cover the table with canvas or hessian.

Roll out coils or ropes of clay about three-quarters of an inch thick. Several coils should be produced at a time and set on one side. Another piece of clay should then be taken, and knocked up into a ball, placed upon the hessian-covered board and rolled out with a wooden rolling pin, or patted down with the flat of the hand to produce a pancake about half an inch thick which will be the base of the form. The pancake should be trimmed with a sharp knife to produce the desired shape and the first coil taken and placed on top of the base around the outside edge. When you have come full circle any excess coil should be nipped off and the two ends joined. Another coil is placed upon the first, ensuring that you do not have a series of joins, one above the other, as this will cause a weakness in the clay form. When several coils have been placed one on top of the other they should be joined together by pulling down some of the clay from the upper coils on to the lower, repeating this all round on the inside and the outside to ensure maximum strength. The surface will inevitably be rather lumpy, but when the coils have dried to a leather-hard state it will be possible to scrape, plane or beat the outer surface into a more satisfactory form.

10 Plaster modelling tools. *Top to bottom:* needle files and a round-back file; riffler and serrated steel tools; spatulated tools; hacksaw blade ground to a knife edge; steel kidney.

This process is repeated as long as the form will support itself. Should it show signs of collapsing, the coiling must be terminated until the lower part of the form has dried. While this drying occurs it is important that the topmost coil is kept damp with a wet cloth covered with polythene, as it is very difficult to join damp clay to dry clay. Even if a satisfactory union is made the damp clay will shrink more than the dry clay, and the piece may crack apart in the final stages of drying, or in the firing. Coiling may be continued until the desired shape is produced, and it is possible to increase or decrease the diameter of the form by placing the coils on the outer edge of each other to widen the form, or on the inner edge to narrow it. This method, and variations of it, can be used to produce very large and/or sophisticated forms. If large objects are to be made the coils should be proportionately thicker.

SLABBING

Where coiling produces rounded forms, slabbing can be used to produce angular objects. Suitable clay is rolled out on hessian- or polythene-covered boards, with guide battens fastened to the edges of the board. The rolling pin should be of sufficient length to enable it to run on both battens simultaneously. By a combination of rolling and trimming the clay it is possible to produce a pancake of even thickness, determined by the thickness of the battens. Several pancakes may be produced and cut to shapes suitable for the fabrication of the form you wish to make. These pieces of clay should be allowed to dry to a leather-hard stage if precise flat planes are required. The pancake of clay, which is to form the base of the object, should then be placed on a suitable board, and where it will make contact with other clay slabs the surface should be scored by cross-hatching, then coated with a slurry made up of the same clay as that used for the body of the object.

The next piece which is to be joined to the base should be similarly scored on that part of its surface or edge where it will join on to the base. By placing this slab on the cross-hatched and slurried part of the base a sound union can be formed between the two pieces by moving them backwards and forwards against each other, which has the effect of bedding them down. While only the first slab is in place it will remain rather precarious, but as soon as the second slab is joined to the first and the base, the object should start to become more secure. This process of assembly is continued until the form is complete.

As in the case of coiling, should the object show any sign of collapse assembly should cease until the lower part of the clay has dried sufficiently to be able to support the subsequent slabs. Each of the joins may be strengthened by making a thin coil of clay, pushing this home down the length of the join, and smoothing it into the surrounding slabs. This strengthening process may be carried out on either the inside or the outside of the joins, or both. You should be most careful to ensure that there is little or no evidence of a join between one plane and another, and while the piece is drying you should inspect it from time to time to make sure that it is not cracking. If any cracks do occur they should be filled with clay of a similar dampness. If you add soft clay to clay which has already dried to some extent, then the cracking will recur.

PRESSING

Clay may be pressed into or on to other objects. This process is normally carried out either over a hump mould of plaster, wood or bisquited clay, or into a plaster mould. If the clay is to be pressed over an object it should be rolled out, as in slabbing, to an even pancake. You should measure the size of the object, and the pancake of clay should be at least as large as the surface on which it is to be pressed. While the clay is still soft it should be taken from the laying-out board together with the canvas and placed on top of the mould. If the mould is small enough it is more convenient to place the mould upside-down on the pancake and then turn board, canvas, clay and mould the right way up. The board is then removed and the clay eased down on to the mould using a sponge and water. When this is complete the hessian may be removed and the excess trimmed off the pancake, which will have taken on the form of the mould.

Where clay is being used to press into a mould it may be rolled out into a pancake if the mould is not too deep. Otherwise, fairly soft clay should be taken and pressed home piecemeal into the mould. The next piece of clay should be laid into the mould in such a way that it overlaps the previous layers of clay to ensure a sound union between the several pieces. If the mould is complex or has considerable detail on the moulding surface it may be necessary to cover the inside with a slurry of clay, and then back this up with plastic clay pressed well home. When the complete object has been moulded the interior of the form should be modelled to a smooth state in order to avoid cracking. This smoothing process should be carried out when the clay is still in the mould. As the clay dries it will shrink and release itself from the mould, unless there are undercuts of sufficient depth to cause the object to lock in place. If this happens it may be necessary to fabricate those areas of the mould, including the undercut, as separate pieces and assemble them when the clay has been removed from the mould.

The quality of the outer surface of the pressed form will only be seen when the clay has been taken from the mould, and any evidence of the method of pressing, such as cracks and fissures where one piece has been pressed against another, should be removed by modelling, unless it is desired that they remain as a feature of the object. (Leaving such creases is rather dangerous as they can be the source of cracking if the clay has not been joined together well enough.) If the clay has been very soft when pressed into the mould it may be that the smoothing and pressing action of moulding is sufficient to form a good union between the several pieces of clay.

CASTING

Clay may be cast into a plaster of Paris mould if it is mixed with sufficient water to bring it to a liquid state. The mould must be watertight and the liquid clay or slip is poured into the mould until it is full. The plaster will absorb moisture from the slip, causing a dense layer of clay to form against the plaster surface. When this has reached the desired thickness the mould may be drained and the clay skin will be left adhering to the plaster surface. After some time this cast clay will dry and shrink away from the mould, when it may be removed. Moulds may be made in several pieces so that quite complex forms can be cast. The plaster forming the mould walls should be of an equal thickness throughout as the thickness of the plaster plays an important part in determining the speed of casting and the thickness of the cast clay in any one area.

If a simple clay and water suspension is used, the mould will quickly become saturated with water and the cast will shrink excessively as it dries out. In order to make casting easier and faster, and to remove the need for over-sized moulds, a deflocculated slip may be used.

Deflocculation

Clay and water suspensions tend to settle out the clay very quickly, and the suspension must be constantly agitated to avoid casts which are thick at the bottom and thin at the top. A deflocculant, or electrolyte, has the effect of changing the electrical charge in the particles of clay so that instead of tending to stick together, a proportion of them will repel each other. This has the effect of dramatically reducing the amount of water required to produce a smooth-flowing slip. The proportion of deflocculant to clay and water must be carefully calculated, but several suppliers of suitable bodies will specify the amount of deflocculant and water required to render the clay to a satisfactory casting slip. The commonest deflocculants are soda ash and sodium silicate, together or separately, but these two together are unlikely to exceed one per cent of the total weight of clay in the recipe. The normal procedure for making a slip is to put the required amount of water into a blunger, together with the deflocculant already dissolved in a small amount of water. It may be necessary to heat this water to be sure that all the deflocculant is dissolved. This solution is then added to the water in the blunger and the clay, in either plastic or powdered state, is slowly added to the water while the blunger is switched on. The resultant slip will be seen to flow as a creamy solution and should be passed through an 80-mesh sieve before use.

Deflocculated slips properly formulated normally contain around 30 per cent of water. Without a deflocculant this would be barely enough to render the clay plastic.

These descriptions of basic methods of forming clay are by no means exhaustive: the inclusion of such very brief summaries of these techniques is to indicate to the beginner the kind of procedures involved in making ceramic forms, so that later chapters may be better understood.

4 Kilns

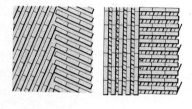

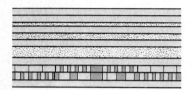

11 *Top left:* First layer of bricks set in a clamp (seen from above).
Top right: First course with second laid over. *Bottom:* Section through bricks set in a clamp. The bottom two layers of bricks are set as above; the spaces between are filled with coal dust or cinders and then bricked over, after which the layers are alternately bricks and coal dust.

One of the simplest types of kiln, originally used for firing bricks, is called a 'clamp'. There are several variations, but the crudest consists of a chamber without a top, built initially of unfired bricks, into which the production bricks are packed, sometimes incorporating layers of straw between the layers of brick. The stacks of bricks are not necessarily continuous within the clamp, but may be placed to leave passages through which the hot burning gases may move, or the spaces may be filled with fuel, normally wood or coal, which will burn when the temperature in that area is high enough. Stoking holes, or 'fire mouths', are constructed around the side walls of the kiln so that when the clamp has been packed, topped with more bricks, and partially sealed with clay and earth, the fire may be lit and the clamp fired. The firing might last up to several days, depending on the number of bricks in the clamp. Once the firing has started the clamp can be left to burn itself out.

The bricks produced will vary in colour and hardness according to how near or far they are from the fire mouth. In order to counter this effect, or at least improve the chances of the interior bricks being fired to a temperature where they would be ceramic rather than dry clay, coal dust or breeze (small fine cinders) used to be added to the clay.

When the clamp has cooled to the point where the bricks can be removed it may be left standing ready to fire the next batch of the production, because the raw, unfired bricks used for the initial construction have been fired hard enough for the clamp to be re-used. Such clamps can last two or three years before they have to be stripped down and rebuilt, this time with fired bricks. Sometimes the brickmakers move to another site before the clamp becomes unusable.

A more sophisticated type of clamp, known as a 'Scottish kiln', is constructed by building two parallel walls with fire mouths at regular intervals. Earth is then banked against the walls to buttress them and improve the insulation. The green bricks are packed in the kiln and both ends closed up with several layers of bricks. One end may be bricked up in such a way as to provide a stoking hole, if the kiln is short enough for there to be no need for side mouths. The kiln is topped off with several layers of fired bricks, and through this topping the smoke is allowed to rise when the firing is under way. Wood or coal may be used as the fuel in any type of clamp.

Clamps are up-draught kilns of the simplest type and do not have the advantage of chimneys to draw off the unburnt gases, which are present in some degree in any unsophisticated solid-fuel firing. This is one reason why the quality of bricks fired in this way would be infinitely variable. Variations in colour and texture can add considerably to the visual appeal of structures made from clamp-fired bricks, but some may not resist normal weathering, while others may be weakened by over-reduction or too fast a firing.

Two more efficient and permanent types of kiln will be found more fully described in my *Manual of Pottery and Ceramics*: these are 'up-draught' and 'down-draught' intermittent kilns. They may be fired by gas or oil, but the commonest fuel in large-scale and brick production is coal. The down-draught kiln is more efficient in burning the fuel, and where this type of kiln is used in large works the flues may be used to heat the drying stacks of bricks. At ground level the flues extend through sheds in which the bricks are stacked. Circular kilns, either up- or down-draught, allow for several fire mouths to be set around the base. This arrangement permits the construction of kilns much larger in diameter than is possible when there is one stoking hole only. Within the kiln chamber a bag wall is constructed which runs parallel to the kiln wall and about 18 inches away from it. This deflects the flames as they enter the chamber and prevents the stacks nearest to the fire holes from being over-fired.

Any type of intermittent kiln, that is to say one which is packed, fired, cooled and unpacked, is relatively inefficient. Probably no more than 40 per cent of the heat generated is used to fire the clay within the kiln. The rest is used in

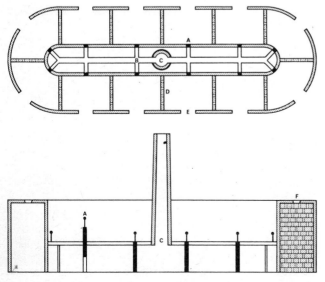

12 *Top:* Plan of a Hoffmann kiln. *Bottom:* Section through a Hoffmann-type kiln. A damper; B flue; C chimney; D chequered brick wall; E kiln door; F top hole for stoking.

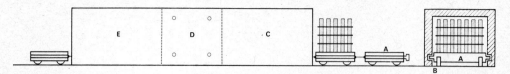

13 Side view and end view of a
tunnel kiln. A refractory trolley;
B sand seal; C heating section;
D firing section; E cooling section.

heating the kiln itself, radiating or escaping from it in the
form of hot gases. However, if the firing changes the clay
object so substantially that the value is greatly increased,
this inefficiency is less important than in the manufacture of
bricks. As most of the heat wasted is taken up in heating the
kiln any type of continuous kiln is significantly cheaper to
operate when there is a constant flow of products to be fired.

The first type of continuous kiln was designed by
Friedrich Hoffmann in 1863, and consisted of several
chambers linked by flues. In large brickworks a Hoffmann
kiln may consist of forty or more chambers arranged so
that they lie as a continuous run, in the form of a circle or
an oval. This type of continuous-chamber kiln enables the
fire to be moved in sequence so that the chambers may be
packed, fired and unpacked without the fire ever going out.
This is achieved by building chequered brick walls between
each chamber. The flues are linked to a central flue and
chimney, with dampers located so that the flow of hot air
from the chambers being fired, or cooling down, may be
diverted to those filled with bricks which are being dried
out, or are in early stages of being fired. As the combustible
material in the brick clay, or the fuel packed into the
chamber, becomes heated the fire is let in. It is controlled
by sealing the trace holes with sheets of paper before packing
the chamber. As the temperature increases the paper
becomes very dry and crisp. When the correct temperature
is reached the paper is set alight and the flames are drawn
through the wall to start the burning in that chamber. As the
firing continues, coal is passed into the firing chamber
through holes in the roof or crown (this is known as top
firing). Hot air is drawn into the chamber from that next
to it which was fired previously, thus cooling that chamber
while contributing preheated secondary air, which improves
the burning efficiency of the fuel in the firing chamber.

The amount of fuelling the fire needs, and its timing, will
depend upon the amount of organic material present in the
clay. Some clays require to be set with coal before the fire
reaches the chamber. Others require topping up to enable
the firing to reach the correct temperature for the desired
quality of brick. The traditional practice of adding coal or
breeze to the clay before moulding encourages good firing
throughout the brick. This self-firing quality is naturally
present in the Lower Oxford clays used in the Bedfordshire
brickworks. The carbonaceous inclusions permit the bricks
to be fired using a third of the fuel considered necessary
when burning other clays.

Tunnel kilns are used to fire sanitary ware, although they
have not ousted the Hoffmann type kiln for large-scale
production of bricks. As the name implies, this type of kiln

consists of a tunnel through which the objects pass at a predetermined rate. As they progress through the kiln they enter a drying and preheating stage where the temperature of the clay is raised under precise control. They proceed through the firing range and then to a cooling section. The speed at which the clay encounters these various zones in the tunnel is regulated so that maximum speed is maintained, but at those temperatures where the clay is susceptible to changes of crystal structure, or other difficult phases in the firing process, the increase in temperature is very gradual. The kiln does not vary in speed but the distance allotted to the increase or decrease in temperature will vary along the length of the tunnel.

The tunnel kiln has many advantages over other types of kiln in that the degree of control is considerably greater, and the heat loss is low as the firing is continuous. The fabric is not constantly being heated up and cooled down, which usually leads to cracks somewhere in the structure of intermittent kilns. Tunnel kilns may be fired by gas, oil or coal. The rate of production from making to finished state can be much shorter than when using other kilns, so that much less storage space is required. The disadvantage is that they are expensive to construct. This may be outweighed by the increased efficiency of the system, but any break in production can leave a kiln running and empty. Closing down the kiln may take several days if it is to be achieved without damage due to fast contraction caused by rapid cooling. Similarly, it takes a long time for a tunnel kiln to reach temperature from cold.

When coal is used to fire bricks in tunnel kilns the top firing technique of the Hoffmann kiln is used and the coal is fed directly on to the bricks from the holes in the crown. Much of the development of tunnel kilns has been carried out by kiln manufacturers in the United States.

Recent developments in the design of burners and insulating materials reveal possibilities for more sophisticated clamp-style kilns which are particularly valuable when considering how to fire a large sculptural piece of ceramic. Sometimes it may be possible to find a friendly brick manufacturer who has spare kiln capacity. In such a case it might be advisable to build the work on his site and use the handling facilities he may have to move the piece into and out of the kiln. Otherwise you could consider the construction of a once-only kiln. The piece to be fired is constructed upon a fireclay or refractory base. When it is complete the kiln is built around the sculpture.

Suitable materials for this type of structure are: lightweight insulating brick, which may be cut easily with a saw to fire any awkward shapes; firebrick for those areas which will be exposed to the fiercest flame from the burners; and ceramic fibre insulating blanket which has unique insulating properties. Depending upon the dimensions of the piece to be fired the kiln may be designed with one or more burners for gas or oil fuels. Provided that more heat can be put into the kiln than is lost in radiation through the walls or convection through a flue or chimney, the temperature within the kiln will rise. The higher the final temperature is to be, the more efficient must be the insulation, and the more powerful or numerous the burners. It is unlikely that more than four burners would be needed unless they are inefficient or the kiln is poorly insulated.

Each kiln will have its own characteristics which must be taken into consideration. The volume of clay to be fired is a major factor, as are the proportions and thickness of the walls of the kiln when constructed. The insulation should not be more than two thicknesses of insulating brick with, if possible, one or more layers of insulating blanket or board between them. This material is very expensive but if handled with care can be used indefinitely. A loose wool of the Fibrefax type is cheaper and can be used to fill gaps in the brickwork, thus avoiding heat loss, or cold air being drawn into the kiln.

The insulating brick is laid around the sculpture, leaving holes for the burners. As the courses of brick are built up you should avoid joints coming one above the other. If you can afford it, lay ceramic fibre blanket over the first layer of bricks and construct the second thickness of brick to cover gaps left in the first layer, so that the passage of air into the kiln is impeded. If the gaps can be filled with Fibrefax wool, so much the better. Finally, cover the whole construction with mud or clay, leaving space for the burners to be installed. Also allow a secondary air inlet, with close-fitting plug, and a hole at the top to allow any steam or unburned gases to escape. The secondary air inlet is important as it ensures that the atmosphere in the chamber is fully oxidized. If you require reduction conditions during some part of the firing the vent can be closed.

With some experience you should be able to design the flue and chimney so that the kiln operates on the down-draught principle, thus achieving greater efficiency and the possibility of reaching higher temperatures. Some method of preheating the secondary air will also improve the efficiency with which the fuel is burned.

Building and operating this type of kiln is not for the novice. The dangers are considerable. If inexpertly constructed the kiln could collapse during firing, with terrifying results. If the burners are carelessly handled there may be a build-up of unburnt gases within the kiln, and when the burner is lit the whole kiln will explode, hurling hot bricks in all directions. If fired too fast the clay piece may explode as the steam will not have had time to escape from the clay. With a lightly constructed kiln, such as described above, such a 'water blow' could collapse the kiln.

Nevertheless, kilns of this type do bring a degree of freedom and improvisation to ceramics, in keeping with much ceramic art, and firing can be great fun!

Brick chimneys, Hampton
Court Palace. Early 16th
century.

A 16th-century town hall,
now in the Singleton Open-
Air Museum, Sussex. Timber
frame and herringbone brick
infill.

'Adoration of the Magi' by Luca della Robbia. Polychrome faience, fired
in sections.

14 Moving-hearth down-draught gas kiln.

A safer kiln of this type, but only capable of much lower temperatures, is the sawdust kiln. This is an open-topped clamp with no air holes or fire boxes. The bottom of the kiln is filled with sawdust to a depth of about 6 inches (15 cm). The piece to be fired is laid in the kiln, more saw-dust spread around and on top of it, and the whole covered with a metal or refractory material. The kiln will have the appearance of a brick tank filled with sawdust. The sawdust is set alight by using a fire lighter, burning charcoal, or a gas jet, and will burn down slowly, consuming the wood and increasing the temperature to a probable maximum of $1560°–1650°$ F ($850–900°$ C) according to the size of the kiln, the weather (this process must be carried out outside) and the amount of air seeping into the sawdust through the brick walls. If the sawdust does not burn it may be due to lack of air through the brick kiln wall. The kiln may take 24 hours or more to fire and cool. When removed, the piece will be blackened where the sawdust has burned against it. This will not be an all-over patina but a random effect. A more durable object may be achieved by firing the clay to $2000°$ F ($1100°$ C) before subjecting it to a sawdust firing. In this case the sawdust may be regarded as a surface treatment rather than a hardening by heat treatment. If bisquited higher than $1100°$ C some clays will not take up the carbon from the sawdust firing and the patina will not be a permanent feature.

FIRING ATMOSPHERES

The atmosphere within a kiln during the firing will vary depending on the type of kiln and the fuel used to create the necessary heat. The heat may be radiated on to the clay objects, as in the case of electric kilns, or convected, or conducted. Provided that the firing is slow enough the load should be evenly fired. Variations in temperature may be attributed to firing too fast, not allowing the heat to distribute evenly through the kiln, failure of one or more elements or burners, loss of heat due to cracks in the kiln structure, or draughts from ill-fitting doors.

Electric kilns are considered to fire with a neutral atmosphere as the elements do not consume oxygen in the kiln nor do they need an excess of oxygen in order to become hot.

Gas kilns may be of the up-draught or down-draught type, with or without muffles. If the ware is protected by a muffle the atmosphere will be neutral. If the ware is open to the effect of the burning gases the atmosphere may be reduced (oxygen-depleted) or oxidized (more oxygen coming into the atmosphere than is consumed by the burning fuel). The gas may be town (coal) gas, natural gas, bottled (propane) gas or producer gas (own gas-making plant). Producer gas is rarely available outside the USA.

Normally the atmosphere within the kiln can be controlled by keeping open the secondary air vents to ensure an oxidizing atmosphere.

Oil-fired kilns are similar to gas kilns in the varieties available and the control of atmosphere. The main difference is in the type of burners which are necessary to atomize the oil and mix it with air to ensure good combustion.

Solid-fuel kilns may be wood-, coal- or coke-fired, or may burn a mixture of materials, provided they will produce enough heat: cow dung, for instance, is used in countries which have a plentiful supply. Modern design allows the construction of solid-fuel kilns in which the atmosphere may be easily controlled. Traditionally they burned fuel inefficiently, so that the atmosphere tended to be reduced. This may have been relieved by the clearing periods when the fire burned down before refuelling, but maintaining a satisfactory proportion of oxygen proved to be so uncertain that expensive or fragile objects were packed in saggars to defend them from the variations in atmospheric conditions.

The effect of reduced atmospheres upon the ware may create different faults according to the type of production. The most immediate effect is that it causes the metal oxides, present either in the body, the engobe, or the glaze, to change their type. For example, copper normally provides a green colour as it takes the form of cupric oxide, but when fired in reduced conditions it produces red colours, as it is converted to cuprous oxide. Iron oxide (ferric oxide) will be converted to ferrous oxide, which is a very active flux.

Reduction conditions may be induced in electric kilns by the introduction of carboniferous material. Providing this is done with care and in not more than one firing in four, so that the oxide coating which is built up, and protects the elements, may be replaced, the working life of the element should not be appreciably shortened.

Controlled reduction for colour changes is difficult. The exact amount and time must be determined by experiment with the kiln and fuel concerned. In gas kilns it is easier to reduce as the temperature rises, but any reduction will impede the progress of the firing and demand a fuel consumption greater than firing to a similar temperature under oxidizing conditions, which is a more efficient use of the fuel. Reduction may be achieved by closing the secondary air vents, and closing, but not shutting, the damper. If the temperature fails to rise it may be necessary to open the secondary air vent a little, or to alternate between reduction and oxidation until the final temperature is reached.

Matured bodies and glazes are not dependent upon temperature alone but upon the amount of heat work to which the clay and glaze have been subjected. This heat work is the ratio between the time and the temperature. Slowing down the firing by reduction may produce over-fired results if the temperature reached is that of a faster oxidizing firing.

15 Stacking unfired tiles on refractory cars in preparation for bisquit firing.

16 Floor and ridge tiles set in a down-draught circular kiln.

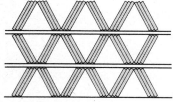

17 Roofing and floor tiles set with layers of fired tiles between each two layers of unfired.

18 Setting decorated and glazed tiles in tile cranks prior to glost firing. The tiles are delivered by conveyer (bottom left).

Kilns are packed according to the type of objects being fired. Bricks and unglazed tiles are packed one upon the other as they are less likely than most objects to stick together during the firing. Spaces must be left to allow the hot gases to flow amongst the stacks or 'bungs' and the clay pieces must be placed so that they will not collapse or distort with the weight of subsequent layers. Roofing tiles, for example, are stacked upon their edges in groups of two or four leaning against each other so that they do not fall over. Sometimes each layer is covered with old tiles so that the weight of subsequent layers is evenly distributed. If the first layer is set parallel to the mouth of the kiln the next layer will be at right angles to it and so on. In this way the flames and heat from the fire will spread throughout the kiln along these channels. Bricks are similarly packed.

Wall tiles are set upon each other for the bisquit firing but the stacks are interrupted by small kiln shelves so that unevenness in any one tile will not affect more than five or six others. Glazed tiles are set in tile cranks which keep the tiles apart and ensure an even distribution of heat across the glazed surface.

Bricks and thicker tiles which are glazed need not be set with special kiln furniture if there is no chance of the glaze moving during firing. If there is any chance of this happening the pieces should be set upon small pads of clay so that no piece will touch another.

HEAT MEASUREMENT IN KILNS

The temperature within any kiln may be measured by means of one or more thermocouples linked to galvanometers. This will measure the temperature of the kiln space at the tip of the thermocouple but is no guarantee of the temperature throughout the kiln. Some kilns have several probes set at various places but draw trials are still relied upon to check the progress of the firing, often in the form of 'Buller's rings', which shrink as the temperature is increased. These have the advantage over pyrometric cones as they will cover a wide range of temperatures, whereas the cone is formulated to slump within a temperature range of 10° or 20°. The rings are placed throughout the kiln so that some may be withdrawn through convenient spyholes while others can only be reached when the kiln load is unpacked. Even these will be useful as they indicate the evenness or otherwise of the firing and remedial action may be taken if necessary in subsequent firings.

Prior to these inventions the fireman would judge the progress of the firing by eye and experience. At the Coade works, at the turn of the eighteenth century, progress was measured by means of a clay strip set in the kiln; this could

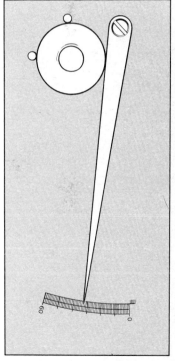

19 Buller's ring gauge.

be inspected periodically and the shrinkage checked against a graduated scale to given an indication of the temperature/heat work within the kiln.

SHRINKAGE TEST

From time to time it will be necessary to calculate the shrinkage of clay so that the object being made will be of the correct size when fired. There are two types of shrinkage to be considered: that which occurs between making and drying ('drying shrinkage'), and that which occurs between making and finishing. The drying shrinkage will be important if you are to make something larger than the kiln on the assumption that it will shrink during drying and fit into the chamber. Moulds for pressing and casting must be made from models which are accurately scaled up to permit the mould to produce a cast which will shrink to the correct size after firing.

The clay pieces for the shrinkage test should be pressed out so that they may dry and fire without distortion. Each piece should be about 20 cm long and 5 cm wide. Several pieces should be made, so that the average shrinkage may be arrived at in the event of any variations in the results. Each piece should be clearly marked with a 10-cm line. Every effort should be made to keep the pieces flat during drying and firing. The lines are re-measured at the following stages, and a note made of the measurements:

measurement	
1	plastic state
2	dry state
3	after bisquit firing
4	after final or gloss firing

Shrinkage as a percentage is calculated thus:

plastic to dry $\qquad \dfrac{1-2}{10} \times 100\%$

plastic to finish $\qquad \dfrac{1-4}{10} \times 100\%$

5 Firing faults

Some of the faults here are not necessarily the results of bad firing but may only be evident after the clay has been fired and therefore must be considered at this' stage. A more detailed description of the chemical and physical changes which occur during the firing of ceramic objects may be found in my *Manual of Pottery and Ceramics*.

Blowing

The clay is completely shattered because the temperature was raised too quickly in the initial stages of firing. Either open the body by adding sand, grog or sawdust so that the clay may more easily give up the moisture essential for its formation, or slow down the early stages of the firing cycle. The thicker the clay wall of the object the more slowly it must be dried and fired. If drying facilities are available, leave the clay form in the drier for a longer period.

Splitting and dunting

The clay object when taken from the kiln proves to be cracked. This may be caused by splitting (cracking when the clay is being heated) or by dunting (cracking when the clay is cooling). If the piece is glazed the crack will show signs of the glaze melting after cracking but if it is dunted the glaze will show a very sharp break. Change the firing cycle so that the temperatures at which the silica conversions take place are passed very slowly, or reduce the amount of free silica in the body.

Black heart

Upon breaking, the fired clay is found to have a black or dark grey interior which has diminished the structural strength of the ceramic. This is caused by firing too rapidly in the early stages of the firing cycle, or to the clay body being too finely grained to allow the water to evaporate through the capillaries in the clay. Water has a reducing effect if trapped in the clay and heated to 705°F (374°C). Any iron in the clay will be reduced and may not re-oxidize as the air will not easily reach into the interior of the clay. At high temperatures the reduced iron will behave as a very active flux and vitrify the clay at temperatures much lower than normal.

Bloating

Blisters form on the surface of the fired clay. This is normally due to over-firing when the clay starts to give off gases, which are not released through the surface. Thus they create unpleasant lumps which can be seen as cavities when the piece is broken. It may also be due to air expanding in the clay during vitrification, or when the glaze has just melted and prevents the air from being released.

Warpage

The form is found to have gone out of shape. This is due either to over-firing, when the clay has become so soft that it can no longer support its own shape, or the piece has been placed in such a way in the kiln that it has been subjected to uneven firing at some point in the firing cycle. If the former is the case, either fire the pieces to a lower temperature in the future or provide a setter for the shape. If the latter fault occurs, avoid placing pieces in that part of the kiln.

Spit-out

1 Reduced iron particles in the clay in the form of iron sulphide may react with water to produce gases which dislodge part of the surface of the object, particularly during glazing after a reduced first firing.

2 If large particles of calcium are present in the clay they will be dehydrated by the action of the firing. Upon cooling they will take up atmospheric water, expand, and generate sufficient pressure to throw off parts of the face, and in severe cases can split the object apart. The lime (calcium carbonate) can be stabilized if taken up into the clay body by melting into the glass which is formed in the clay during the firing. The effect of this integration is to modify the colour of red-firing clays to buff or yellow, depending on the ratio of calcium carbonate to iron oxide in the unfired clay; and the temperature, or degree of vitrification reached during the firing, and the length of time taken for the clay to reach that temperature.

The standard test to check, not the amount of calcium present in the body, but the proportion which may cause blow-outs, is to weigh a small amount of clay and mix it with water so that it may be passed through a 30-mesh sieve. The residue is treated with dilute hydrochloric acid which will dissolve the calcium carbonate. The loss in weight of the clay, including that which has been treated with acid, and the filtered or dried out clay, indicates the amount of lime likely to cause blow-out. Any which will pass through the sieve will be taken up into the clay during the firing.

Efflorescence

Some red-burning clays display an alarming tendency to scum (a white deposit upon the fired surface). This is

particularly offensive in terracotta modelling and is objec-
tionable in brickwork. It is caused by calcium or magnesium
sulphates which are soluble if not integrated with the clay
body. Rather than remake a piece of sculpture it may be
possible to refire the piece to at least 2080° F (1140° C) and
soak it for about half an hour, which will allow the sulphates
to react with the clay and become insoluble. Once the
scumming has occurred, any attempt to wash it off will
only dissolve it, and it will be re-absorbed into the brick.
The proper cure is prevention, by adding barium carbonate
to the clay which will convert the magnesium and calcium
sulphates to carbonates. The barium becomes barium sul-
phate, which is insoluble and therefore causes no problem.

6 Bricks, roofing tiles and pipes

For many thousands of years bricks have been an essential architectural feature. In those areas where there is little natural building material, bricks of some kind can usually be manufactured or, if imported, at least have the joint virtues of coming in a size and shape which are easy to handle, and of not being subject to environmental degrading unless they are unfired. The fired strength of a brick will vary according to the clay, the method by which it is made and the heat work to which it is subjected. Even the softest or easiest fired brick will find a use in the building of smaller houses where the compressive strength will be enough to withstand many times the weight which it will need to carry. The appearance of bricks may vary in colour and texture, and with the use of inexpensive stains this variety may be further increased.

At present there are three major types of brick which are easily available: common bricks, facing bricks and engineering bricks. Some common bricks may be suitable for the construction of exterior walls, but facing bricks are made specifically for this purpose. Engineering bricks are much harder than the other types, having a considerably reduced porosity – less than 2 per cent – and are suitable for special uses and situations, such as surfaces which are subjected to acid attack, or very heavy wear. Engineering bricks are used in some European countries to pave roads where the initial high cost is more than compensated for by the small amount of maintenance required. Such bricks are known as paviors and may be half the thickness of normal bricks.

BRICKMAKING CLAYS

Clays for brickmaking vary in type and quality as there is no single criterion which can be applied when determining whether or not the clay deposit is suitable for brick production. Because bricks are required in large quantities production is only viable if the deposit requires no drastic modification other than the crushing of all the inclusions to permit the clay to pass through the making procedure without blocking the equipment. A convenient source of fuel is essential if production is to be economic.

The earliest known bricks were not fired but were a mixture of clay and sand reinforced with straw. The clay/sand mixture was probably a natural deposition from the uplifted bed of a lake or river. The straw, which was added during the making process, would have a similar effect to that of fibreglass in glass reinforced plastics today. Naturally unfired clay bricks, even when reinforced in this way, will not last very long in a climate with periods of heavy rainfall. Where such building methods are used today the structures rarely survive more than two rainy seasons without serious weathering, and if unrepaired will reduce to a muddy deposit. During this process of erosion unfired bricks gradually lose their structural strength and are considered unsuitable for the construction of load-bearing walls more than twelve feet high. All bricks which are to be fired must be made of a clay which includes some fluxing material and the most common of these is calcium. Many alluvial clays include calcium of very fine particle size, and clays which have formed on a sea bed may include calcium in the form of shells. These can sometimes be seen when the bricks are broken in half as, even during the firing, some of them do not lose their shape. This calcium acts upon the alumina and silica present in the clay to produce a strong and dense brick at an economic temperature of about 1830° F (1000° C).

Siting of clay brickmaking factories

As the value of clay is not substantially increased by its conversion into bricks, the brickmaking factory, whether it uses hand processes or machinery or a combination of both, must be adjacent to or even on the clay site. The cost of transporting large quantities of clay over long distances would greatly increase the cost of the finished brick. The clay is dug or quarried from the deposit according to its state. If the clay is hard and dry it may be quarried like stone, loaded into containers, and transported to a crushing plant where the particle size is reduced to that which is most convenient and economic for the production of bricks. If the clay is in a plastic state, it may be washed out using hoses, and run off into settling beds where it is allowed to dry, and dug from the beds as required.

It is not normal practice for the clay used in brickmaking to be modified by the addition of any material such as colouring oxides, or the removal of material naturally present in the deposit. If the clay will not produce a satisfactory brick without too much modification it may be regarded as uneconomic for brick production. Smaller brickmaking plants specializing in a more exclusive market may find it possible to prepare the clay with more care. Usually when the good working seam has been exhausted the brickmaking plant will be closed down and moved to another and more attractive site. However, clays from two or more deposits may be combined where they contribute desirable characteristics and the precise proportion of each clay present in the brick is not so critical that the mixing has to be carefully controlled.

TYPES OF BRICK

Bricks may be classified by their colour, their mechanical strength, and the purpose for which they are intended. Common bricks used in building where their appearance is of little or no consequence may vary in colour from red to pale yellow. They are not vitrified and are not normally fired higher than 1830° F(1000° C). They are not expected to have a very great compressive strength. Bricks which are to be used as facing bricks are more carefully made and the quality of the clay is more rigorously controlled. They are sometimes fired beyond 1000° C to produce a more acceptable surface and this is accompanied by greater mechanical strength. Engineering bricks and paving bricks are fired to the point of vitrification or to a temperature approaching this state under reduction conditions, which achieves maximum strength without deformation of the brick. This reduction normally produces a blue-black brick because the iron present in the clay reacts with the other minerals present. Such bricks have very great compressive strength and the production is carefully controlled. With both facing bricks and engineering bricks the firing schedule often includes a long soaking period to allow the clay to mature fully, without the risk of 'bloating', or deformation.

If sufficient calcium is present in the clay mixture it will bleach the iron present in the body and produce a yellow or dark orange fired colour rather than the expected dull red.

Should the calcium be of sufficient particle size it will not fuse completely with the clay minerals, but remain as dehydrated calcium within the body of the brick. It is most likely that after some time, perhaps several years, this dehydrated calcium will absorb atmospheric moisture, causing it to expand and the brick to throw off part of its finished surface, leaving unsightly craters. If this process continues it will substantially weaken the mechanical strength of the brick. This fault in bricks, known as 'spit-out', can be attributed to calcium present in the clay when a large white particle can be seen at the centre and lowest point of the crater.

Another type of 'spit-out' is caused by large particles of iron in the form of iron pyrites. In this case there will be black slag-like particles at the base of the craters.

BRICKMAKING METHODS

There are four methods of making bricks:
  by hand
  by extrusion
  by plastic pressing
  by dry pressing

20 Wooden bullnose brick mould, producing a brick with a frog on one side (fired example on the right).

Hand moulding

Until the Industrial Revolution all bricks were made by hand, and hand brickmaking existed on a considerable scale well into the twentieth century. The site of the clay deposit would be chosen as near as possible to the site where bricks were to be used. The brickmaker would dig the clay, removing any large stones, and possibly pugging through a horse-driven pug, make his bricks and fire them in a clamp of unfired bricks. If the site became established, he might construct a simple up-draught kiln. The clay might be used immediately or left through the winter months to 'lean'. In the spring it would be dug up, and wedged together with additives, such as sand, to produce a more open texture. This helps bricks to dry. Coal dust or ash, which helps the firing, might be added. The water content of clay must be standardized to regulate the shrinkage of the bricks from the wet to the fired state (about 15 per cent).

To make bricks by hand you need a strong wood or brick table and a mould in the form of a deep wooden frame. This is constructed to produce a brick of the desired dimensions after the shrinkage which will take place as the brick is dried and fired. Beech used to be the most popular

49

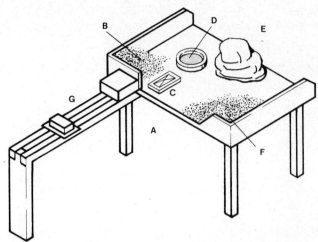

21 Brickmaker's bench.
A brickmaker's position;
B brickmaker's sand; C stock;
D water tub; E clot-maker's position;
F clot-maker's sand; G page, upon
which the moulded bricks are set.

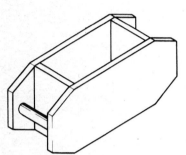

22 Individual brick mould.

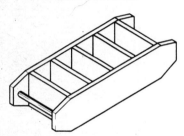

23 Multiple brick mould.

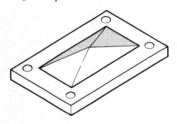

24 Stock to provide a frog in the finished brick.

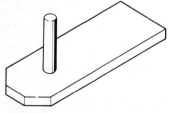

25 Plane for smoothing the top of the brick while in the mould, and for compressing the clay to strengthen the brick.

wood for mould making as it is said to release the clay more easily than other woods. Sometimes the moulds would be reinforced with metal straps on the top edges. The bottom face of the brick is formed by a stock of such dimensions that the mould will fit easily over it. The stock is fastened to the bench and may be flat or include a rectangular pyramid projecting upwards. This will serve to form an indentation, called a 'frog', in the base of the brick. Such a shape is said to strengthen the brick, and it certainly provides a key for the mortar.

The stock may be fixed to provide a standard depth of brick or it may be held in place with long nails, so that the specification can be changed from day to day. The mould and stock are liberally coated with sand between each pressing of clay; this provides a good texture on the bricks as well as facilitating removal from the mould. The clay is kneaded, or wedged, on the moulding table, which should be coated with sand to prevent the clay from sticking. It is knocked into a lump, or 'clot', of roughly the correct size for the mould and then slammed into the mould with enough force for it to fill the corners. The surface is then struck off level with a blade called a 'strike', or with a harp.

After being moulded the clay, still in the mould, is lifted from the stock and jerked, or shaken, from the mould on to a pallet, on which it is carried away from the table and stacked, in such a way that air can circulate around the bricks.

Before the establishment of permanent sites these stacks would be set with layers of straw between the layers of bricks. The stacks, or 'hacks', were eight bricks high and were completely covered with straw to prevent the wind drying the bricks unevenly, or the rain from eroding them. After being stored in this way for three or four weeks the bricks would be ready for firing, but the exact length of time depended upon the prevailing weather. The bricks were then fired in a clamp of unfired, or partially fired, bricks. Sometimes the unfired bricks would be restacked in a herring-bone formation ('scintled') when leather-hard.

This process of brickmaking requires a lot of practice and is very exhausting work, but in the nineteenth century a team of four or five, working under a lean-to, open-sided shed, would average about 300 bricks an hour over a ten- or twelve-hour day. Often a family would form a team, the wife acting as clot-moulder and the husband as the moulder, while one child brought the clay to the table and another loaded the barrow and took the bricks to the stacking area. As all the moulding was in the open air the work was considered seasonal, i.e. for six or eight months of the year.

Hand moulding is a flexible method of brickmaking. Other types of moulds may be used and they may be made from materials other than wood. If a sandy surface is aesthetically undesirable, and if the clay tends to stick in the mould, it may be necessary to treat the surface of the mould with a release agent. Oil or grease may be applied to most mould materials, such as wood or metal, but if the mould is naturally porous, as for instance plaster of Paris, the answer may be to leave the mould to dry.

If the brick is to be anything other than a flat-faced rectangular block it may be necessary to provide a top and bottom face to the mould, In this case less clay will be required to provide a rectangular brick of the same dimensions as a plain one. The mould is assembled without the top face and the clay inserted through the open face. When filled and levelled off, the top face is placed over the brick and pressed home. The finished brick will be more dense than those made simply by filling the mould, and will shrink less than an unpressed brick.

Extruded bricks

Extruded bricks were made possible by the development of pugmills to the point where they could not only shred and mix the clay, but also extrude a dense, integrated mass. The clay must be free from large stones as these would tend to damage the auger which drives the clay down the barrel of the pugmill. The mill itself is normally hopper-fed from a continuous conveyor feed, sometimes with a secondary pan grinder to ensure that no stones which have escaped the action of the main crushing and grinding equipment pass into the pugmill uncrushed. Once inside the mill the clay is shredded and driven towards the die face. Prior to its extrusion the clay is compacted by a reduction in the size of the pugmill barrel.

According to the type of bricks being produced it may be necessary to include in the pugmill a de-airing section, evacuated by a pump so that no air is present, and any carried by the clay is drawn off. This is not the usual practice as the clay must be carefully prepared, which will substantially increase the cost of production.

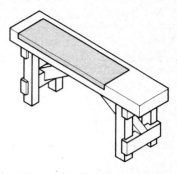

26 Dressing bench, with an iron inlay on which the bricks are rubbed and trued up with a dresser before firing.

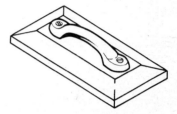

27 Clapper with which bricks are beaten as they lie on the drying floor, to correct any warping.

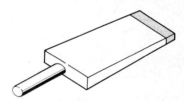

28 Iron-tipped dresser, used to correct any warping in bricks at the leather-hard stage.

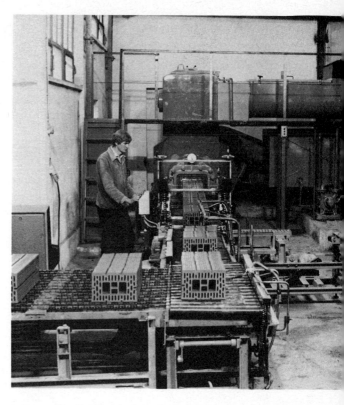

29 Extruding hollow clay blocks.
The system is fed by a mixing auger
(top) leading into a de-airing pugmill
(centre) with a die designed to
produce a continuous extrusion.
This is carried forward on rollers to
the travelling table, from which the
blocks are conveyed to the driers
(not shown).

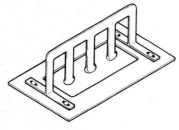

30 Die plate seen from the back.
This type of die does not require a
spider to support the centre cores.
The clay is divided by the back of
the core plate and joins together as it
travels through the face of the die.

Pugmills designed for the production of bricks are
enormous pieces of machinery, and bear little relationship
to those which might be found in a studio. Some are designed
to extrude many bricks at once by having a very wide
extrusion mouth containing several dies. The die is designed
to extrude the shortest face of the brick, so that a continuous
strip of clay exits from the die on to the conveyer system.
The first conveyer is no more than about three feet long,
after which the strip of clay passes on to a travelling table
consisting of a cutting bed with a wire cutter. As the clay
passes on to this bed it triggers the cutting wire, which
moves up and down to cut the clay, as the travelling table
moves forward at the same rate at which the clay is being
extruded, so that the clay is cut vertically. Without this
device the cut would be at an acute angle rather than at
90° to the surface of the brick. When the wire has passed
through the clay it springs up and back ready for the next
cutting stroke. Before this automatic solution was engineered it was necessary to cut off a length of extrusion and
place it upon a cutting frame which, when pressed down
upon the extrusion, would produce bricks of equal length.
Otherwise the extrusion had to stop until the bricks were
cut and moved away.

Some cheaper bricks are wire-cut longitudinally as they
come from the die. In this case the die is shaped to produce
an extrusion conforming to the dimensions of several bricks

placed side by side. The wires are located about 15 to 30 cm away from the die mouth, so that as the clay is extruded from the mill it is forced up against the several wires which cut the wide extrusion into several thinner ones. These then pass to the cutting-bed, as previously described. Wire-cut bricks are rather rough-sided and are normally considered too unsightly to be used for exposed brickwork. The rough finish can, however, be an advantage when rendering with either cement or plaster as it provides a satisfactory key.

Flat-topped bricks do not provide a strong physical key when laid in the normal way with sand and cement mortar. To provide a brick with improved keying properties it is necessary to extrude them on their sides, so that the strip is approximately 20 by 10 cm for standard bricks. It is then possible to design a die which produces an extrusion with holes running horizontally through the longer faces. These holes are normally about 12 mm in diameter, and there may be up to fourteen of them in each brick. When the bricks are laid the mortar is squeezed up into the holes and locks the brick in place once the mortar has set hard.

The die, which determines the outer faces of the brick, is designed in the normal way and is located at the mouth of the pugmill. The rods which produce the round holes may be fastened to a spider which is clamped between two sections of the pugmill barrel so that it lies at right angles to the line of extrusion. The steel rods project forward so that their faces lie on the same plane as that of the outer die or up to 0·4 mm behind it. The clay is divided as it passes through the spider and is compacted by the tapering barrel of the pugmill. In this way it is rejoined around the internal rods and is driven parallel to them, up to the outer die through which it passes to produce a pierced brick.

Pierced bricks can also be produced by more sophisticated dies. Hollow bricks are a natural further development, but the thinner the walls of the brick the more carefully the clay must be prepared to avoid inclusions which would block or damage the dies. In most cases de-airing of the clay will be essential if a high loss rate is to be avoided. De-airing also tends to increase the plasticity of the clay, which will facilitate the extrusion of complex hollow bricks. These are often designed with ribs on all the outer faces, which act as keys for mortar or plaster depending on the face in question. They also help to decrease the chance of deformation while in an unfired state, as they act as corrugations in the clay wall and as buttresses when stacked in the kiln.

As with other methods of brickmaking, the water content of the clay should be controlled in order to maintain constant finished dimensions. This control of water content is even more important in the case of hollow block extrusion, where too little water will produce an unsightly and ragged finish to the outer surface, and too much water will cause

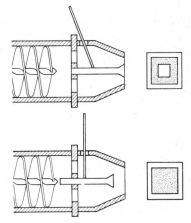

31 Pugmill with movable core. *Top:* Core in forward position, to produce a hollow extrusion. *Bottom:* Core in retarded position, to produce a solid extrusion. By alternating the position of the die, hollow blocks with one or both ends closed may be produced. The core is supported by a spider located behind the mouth of the die. This spider is pierced to allow the clay to progress over the core; the clay is then compacted by the tapered mouth of the pugmill.

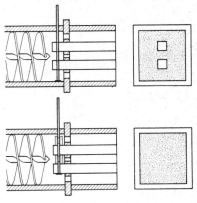

32 Hollow-core pugmill. *Top:* Shutter closed to produce a hollow extrusion. The die is held in place by means of a spider located behind the mouth of the pugmill, and the shutter is operated by moving it up and down.

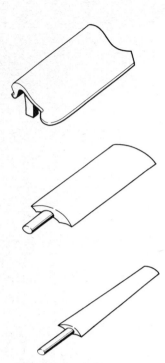

33 *Top:* Washing-off frame, on which tiles are washed into a curved form after they have been pressed into shape by means of a tile frame. *Middle:* Splayer, on which the curved tile is taken from the washing-off frame. *Bottom:* Thwacker, with which the tile is beaten at the leather-hard stage to correct any warping.

34 Thwacking frame, on which the tile is thwacked, or beaten.

35 Thwacking knife, for trimming the tile after thwacking.

the block to deform once it loses the support of the die. Hand pressing may be used to produce hollow blocks which are too complex to be extruded satisfactorily.

Solid bricks which have been extruded may be pressed by automatic presses fed by conveyer systems carrying bricks from the pugmills. For this purpose the extruded and cut bricks may be regarded as little more than blanks. The pressing action of the hydraulic rams compresses and reshapes the clay to produce a brick which is very dense. It will shrink less, as the pore sizes have been reduced by the pressing, and will be dimensionally accurate as the dies are capable of highly engineered specifications. The finished bricks are characterized by a smooth surface, sometimes even with a degree of polish. Most high-quality engineering bricks are produced in this way. The dies are made of machined steel, and must be very hard wearing as the clay is highly abrasive. The presses are normally automatic and may be banked together, up to ten abreast, and fed by one or more pugmills. When pressed, the bricks are ejected from the dies to be carried on conveyers to the kiln or to a drying station.

Plastic pressing

Plastic-pressed bricks are similar to extruded and pressed bricks except that the water content of the clay may be marginally reduced and also the pugmill will feed directly into the moulds. These may be set in a circle so that as each mould is filled by the pugmill it is moved one stage round. As each mould reaches a position opposite the pugmill it is located in the press and the re-shaping is completed by the single action of the press. Some bricks may be pressed several times to improve the accuracy and finish.

Dry pressing

As the name suggests, clay for this purpose has a very low water content. It is commonly used for processing such clays as the shale clay, known as the Lower Oxford clay, found near the village of Fletton in Cambridgeshire. The natural inclusion of oil or bitumen permits the pressing into shape of the granular clay and also serves to lubricate the dies during the pressing action. The clay may be pressed several times (one manufacturer presses four times), with a brief pause between pressings to allow any trapped air to escape. The clay is not pugged before pressing as crushing, grinding and sieving are sufficient.

ROOFING TILES

Roofing tiles are made by a combination of extrusion and plastic pressing. The extrusion is flat, and fed to the dies, which vary in complexity according to the type of tile being made. Previously tiles were made by hand in flat wooden frame moulds called 'patterns', similar to the brick moulds but much shallower. The procedure was exactly like that associated with brickmaking except that the tile, when pressed in the frame, might be re-pressed over a curved mould to provide an arched tile, sometimes designed to interlock when installed.

Ridge tiles were made in the same way but the pressing into shape was done over a shape called a 'horse'. These ridges were made in a wide variety of angles to accommodate various types of roofing structures and junctions.

Decorative finials were hand pressed in plaster moulds as they were based upon models made up in clay. The tile which formed the end of the ridge might be made as a standard fitting but a socket was provided at the end so that one of the several finials could be fixed in place upon installation without the necessity to produce the complete tile and finial in all the varieties shown in the pattern books. Similarly decorative ridge tiles were made with a lengthways recess, so that decorative features could be supplied upon request and be fixed to a standard ridge tile.

36 Tile block and horse. Roofing tiles, after being pressed in a frame, are left to dry and then placed six at a time on the curved tile horse and pressed several times with the tile block, which has the same curvature (or 'set') as the tile.

37 Roofing tile, showing the set (a curve of about 3 metres' radius) and the two 'nibs' by which it may be held in place on the roof.

38 Some decorative finials, from a nineteenth-century catalogue.

55

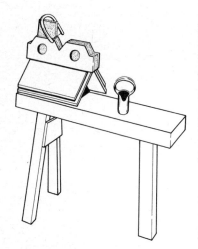

39 Cutting a decorative ridge tile. The extruded or press-moulded tile is set upon a horse and pressed to give the correct ridge angle. The wooden pattern is placed over the upright rib, which is trimmed to shape with a bow. The holes are cut with a circular cutter (centre). The tile, pattern and cutters are well lubricated with oil.

40 *Right:* Stripped-down pugmill showing spider (arrowed) and three sets of dies.

Contemporary ridge tiles are extruded, in some cases by means of a pipe-making pug extruding vertically on to a counterweighted table which is pressed down by the extrusion as it comes from the mill. The mill stops when the tile has been extruded to the correct length, and it is cut from the mill by means of a bow.

DRAIN PIPES

Drain pipes are extruded through de-airing pugmills, usually horizontal, but some are vertical. The pipes are extruded through the floor of the works so that they may be stacked vertically on the floor beneath. Those which are extruded horizontally are forced against a die which expands the end of the pipe so that it will receive a normal pipe end, to form a watertight joint with the help of a bitumen seal. Field drain pipes are not glazed, but sewers and other types of drain pipes are normally glazed, usually with salt glaze. Complex pipe sections, bends, T-junctions and traps are hand-moulded in plaster moulds.

Roofing tiles were traditionally made on the same sites as bricks. Drain pipes are a separate and much later development, and are made by specialist factories.

CHIMNEY POTS

The function of the chimney pot is to prevent the wind forcing rising smoke and gases back down the stack. Formerly, chimney pots were either thrown or pressed in plaster moulds. Hundreds of different designs were produced, each supposedly more efficient than its competitors. Some were more concerned with decoration than efficiency and even the simple thrown pot was thumbed to produce a piecrust type of ridge beneath the roll top. Others were decorated with lines of contrasting clay painted in bands 50–75 mm below the top of the pot. Most pots are between 90 and 120 cm high and might be made in two pieces (one pot placed upon another if they were thrown), then turned around the join to produce a smooth exterior profile.

41 Factory-made chimney pots.

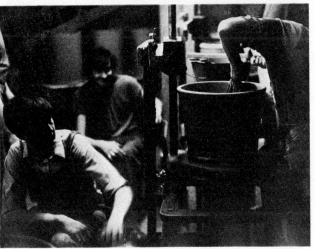

42 Throwing a chimney pot by hand.

7　Faience and terracotta

The terms 'terracotta' and 'faience' are generally understood to mean, respectively, unglazed and glazed earthenware. Contrary to the commonly held belief, 'terracotta' does not refer only to red-firing unglazed clay but may be applied to any unglazed clay. Terracotta has a long history but faience was used on a large scale only during the nineteenth and early twentieth centuries. As building costs rose, and the modern school came to dominate western architecture, the decorative style for which terracotta and faience were most suited fell out of favour; many of the factories ceased to operate on their former scale and diversified or closed. Today, there are very few that can undertake even the restoration work which is now becoming necessary in order to repair the accidental damage to which all buildings are liable.

At least one factory creates a terminological dilemma by calling all hand-pressed facings, whether glazed or unglazed, 'terracotta', and all cast or mechanically made pieces 'faience'. Nevertheless, in describing the processes of manufacture I shall follow the distinction given above.

The clays which are used in this type of manufacturing are usually bodies produced by mixing several natural clays or by adding grog of the same clay, which has the effect of opening the clay without changing the colour of the fired body. Plasticity is not very important but the body should not shrink excessively, otherwise the models and moulds will have to be unnecessarily large. If the piece is to be glazed the colour of the clay may not be considered very important, but the fired colour is crucial for terracotta: not only must it match the design colour but it must remain constant and repeatable throughout the job in hand. Red-burning clays have been by far the most popular for terracotta work – hence the mistaken belief that the term applies to red-burning clays only. It is not uncommon for the red-burning effect to be created by the addition of iron to a buff-burning clay which has the necessary working properties and may be used in the production of faience.

The moulds in which the clay is cast, or more often hand-pressed, are made of plaster although bisquit and wood moulds were used in the past. Models for these moulds are fabricated as described in Chapter 8, and, although only one face of the piece may be revealed when finally installed, the pieces are made in the form of load-

43 This unique example of Doulton terracotta, faience and glazed stoneware was in the Chancery Lane and Holborn branch of the Westminster Bank, London. The building, erected in 1899, was demolished in 1964.

44 Terracotta block pressing table. The moulds to be used are set in a line between the rails on which the table runs. The clay is knocked up (lightly wedged) on the plaster slab on the right, the mould placed on the table, filled with clay and trimmed with the bow (seen lying on the plaster slab). The mould is then set on the floor and the table moved down the line to the next mould.

bearing blocks rather than tile claddings; nevertheless, the term 'faience' is used to describe large glazed tiles. The faces of the blocks which will not show are made so that they will provide a good key for the mortar and one or more holes will be left to facilitate drying and firing. When the piece being pressed is a panel rather than a block, the mould may have an open back so that reinforcing clay ribs may be put in by hand when pressing is complete. Moulds for rectangular blocks with only one decorative face may be made as drop moulds with the back face left open to reduce time – and hence cost – in pressing. Otherwise the back face of the mould will allow for a hole to be left, through which the clay may be pressed.

The clay used for pressing into the moulds should be very soft to reduce the effort needed by the presser. It may be knocked up on a plaster-topped table to harden the outer skin very slightly and force any grog below the surface of the clay. Large or small pieces are then pressed into the mould, depending on the size and detail involved. The clay in contact with the plaster dries quite quickly; the piece can be removed after a few hours if it is in the form of a rectangular block but may need to be left longer if it is more complex. The interior of the pressed clay is finished off with templates to consolidate the clay and to remove the variations in section which can occur when pressing by hand.

45 Trimming a pressed block on a
fixed table. These blocks are similar
to bricks; larger blocks are set on the
plaster slabs which can be seen on
the floor.

46 Trimming a hollow block after it
has been taken from the mould.
Holes are cut in the side to facilitate
drying. The scoring which can be
seen in the side, and which helps the
mortar to grip the block, is modelled
into the mould and reproduced
during the press moulding.

47 Pressing shop in the heyday of terracotta moulding.

Reinforcing ribs may be inserted into large pieces, and in some cases the back piece may be fabricated separately and luted on when it has dried enough to support itself without deforming. Pieces which are to be remodelled after pressing, to introduce more detail or to make additions, can be either removed altogether from the mould or only the face to be altered may be exposed to provide support for the piece if there is a danger of its being deformed.

Sometimes it is necessary to produce a piece which has the function of a cladding, with a form so complex that it is necessary to press it as a block. The parts not required may

48 Press-moulded foundation block for a power station. With this size of block, interior ribs are necessary to prevent the large surfaces from distorting. Sometimes these are joined lightly to the block and knocked out after firing.

62

then be cut away but left in place to act as supports for the faces so that they do not distort during drying and firing. If the clay is soft enough when cut some slight bond should be created, sufficient to hold the pieces in place. Such supports may be left in place to protect the work in transit and only removed by knocking away just before installation.

Any type of faience or terracotta must be carefully dried, and factories still involved with this type of work use either controlled driers in the form of tunnel driers or drying rooms heated with underfloor steam pipes which are regulated to allow the blocks to be dried slowly. This is particularly necessary in the case of blocks where the clay on the inside has very little moving air passing across the faces to carry away the evaporating moisture.

Once the blocks are dry they are fired to the maturing point of the body. If the pieces are to be glazed, the glaze is usually sprayed on with an airograph. Most pieces, either glazed or unglazed, are fired only once. Polychromatic effects are achieved on one firing either by careful design of the modelling so that it holds the various glazes apart, or by applying the colour as underglaze and glazing with a single transparent glaze. Glaze which finds its way into the sides of the block is removed by sponging, so that the adhesion of the mortar during fixing is not affected.

Some very large pieces have been produced for architectural work. The handling involved demands considerable resources of either manpower or machinery. The advantages are that the piece has no unsightly joins and that installation is much quicker than assembling several pieces that are still quite large.

49 Faience façade on a public house (now demolished) in Portsmouth, England. Such façades were supported on a steel frame and glazed in various colours, brown and green being the most popular.

50 The Coade stone tomb of Vice-Admiral Bligh (of the *Bounty*) in St Mary's Churchyard, Lambeth, South London.

High-quality terracotta such as Coade stone seems not to be affected by atmospheric erosion. It also avoids the worst problem of brickwork, in that the amount of mortar is considerably less than in the equivalent brick structure and therefore less repointing is necessary. Cheaper and easier to form than stone, it still has much to recommend it to the discerning architect. Certainly the best terracotta has survived more than 150 years without deteriorating. Coade stone was used to produce large tiles as well as complex sculptural pieces. Some of these were more than 10 cm thick, and made in several pieces fastened together when installed. This material has been the subject of some research but the analysis of the finished material reveals little to indicate why it is so hard-wearing. It seems that the body was made of china clay and grog of the same body together with sand and some glass to act as a flux, although the exact proportions have not been determined. As the Coade family came from Dorset it is possible that they used one particular type of clay; this, together with considerable skill in making, drying and firing, established the Coade works as one of the major London workshops for the production of architectural decoration during the late eighteenth and early nineteenth centuries.

Italy is one of the major centres of terracotta and the façades, details and free-standing pieces demonstrate a richness and application which must have employed not only the skill but the imagination of many artists and craftsmen. Most of the simpler blocks would have been pressed in wooden box moulds but the more complex sections must have been modelled in clay oversize to allow for double shrinkage: the piece modelled as a mould was dried, fired and then used to produce the pieces needed for the building, which in their turn would shrink during drying and firing. The grog included in the clay used for pressing would reduce the shrinkage but the work must have been carried out with considerable care so that the complex geometry of the façade would not be interrupted by a deformed piece of decorative terracotta.

8 Model-making

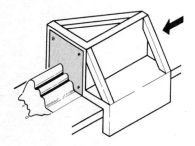

51 Horse, or plough. The template is attached to the front of the plough by four or more screws; it has a bevelled edge. Plaster is run on to the table surface and the plough pushed through it to produce the desired form.

In the production of faience and terracotta it is often necessary to prepare models from which more than one working mould may be produced. Moulds for one-off pieces may be necessary from time to time but this procedure is not particularly complicated as the modelling material is usually clay and the moulded piece can be finished by hand without a great deal of extra effort. Where many pieces are to be pressed the mould should be capable of producing pressings which are as near perfect as possible. Exceptions are those special items such as figures considered worthy of hand finishing rather than using complex moulds which might be difficult to press.

Much of the terracotta employed in the architecture of the late 1890s and the early twentieth century included details which could only be modelled by hand. Nevertheless the fundamental design factor (the installation of several components to produce continuous relief patterns) necessitated an underlying geometry which was the basis for many of the models. Rectangular blocks may be produced by casting within walls or they may be sledged (see p. 67). Circular blocks can be turned on a lathe or by using a horse (see p. 69) if the radius is too great to permit the model to turn without running against the bed of the lathe. Where arcs of a very large radius are required the horse may be several feet long from the centre pivot to the outer circumference. In all cases the models are made from plaster of Paris although some of the earlier pieces may have been of wood. Plaster remains the most convenient material as it is free from laminations or grain which would hinder the modelling of some types of detail. Furthermore the models are made larger than the fired clay pieces will be and are said to be 'model size' as opposed to 'finished size'. This difference may be as much as 10 per cent or more, to allow for shrinkage.

Several types of plaster are available commercially, varying in such qualities as fineness of grain and hardness of setting; for model-making, only the finest-grained types are suitable. The proportion of water to plaster may vary according to type, but manufacturers of these special plasters usually supply technical information such as this upon demand. Whichever type of plaster you decide upon for the kind of model-making you intend to undertake, it must be hard enough to withstand the handling and mould-making which will follow.

SLEDGING AND TURNING

These are terms used to describe methods of forming plaster while it is still soft. The procedures are therefore carried out quite rapidly, before the plaster has had time to set, and some practice will be needed to make satisfactory models. Under no circumstances should you deviate from the recommended ratio of water to plaster, as this may adversely affect the properties of the plaster when set. There are commercial additives which can be used to increase the setting time for some plasters, but the manufacturer's recommendations should be followed in this respect.

Sledging should be carried out on a glass-topped bench. The glass must be strong enough to withstand the weight of the plaster and sledge: 7-mm plate is best, set on a perfectly flat table top. The glass is fastened with screws so that the ground edge overhangs the edge of the bench by about 3 mm along its length. The template which is to be drawn through the plaster is made from some easily cut metal such as zinc (used litho plates are ideal for this purpose). The desired section of the plaster model is cut from the metal and fixed to a wooden backing so that the edge of the metal template may shape the plaster, with the wood providing additional support and rigidity. The template is then fastened to the sledge, and the plaster mixed up and poured in approximately the right position. You must judge the state of the mixture so that it is poured as it starts to gel: if the plaster is too wet it will run into a flat puddle. When properly mixed, it will retain most of its poured shape and when the sledge is drawn over it the plaster will be formed into approximately the correct shape. Most sections will require more than one pouring of plaster to complete the height and length of section required. These subsequent pourings must follow quickly one upon the other to ensure that they bond well and avoid the problem of the model shelling apart in several layers. When sledging very fine sections, the column of plaster may tend to be pushed along the glass, and to overcome this difficulty small nails must be fixed into the glass to act as anchors. Using a fine diamond drill, small holes should be made in the glass without penetrating right through. Headless nails are then fixed into the holes either with plaster or with adhesive, in such a way that they may be removed when the task in hand is complete so that they do not interfere with future work.

When moving the sledge through the plaster the pressure should be applied not only down the length of the glass but also away from you, so that the running edge of the sledge remains at all times in firm contact with the polished edge of the glass. Any irregularities in the glass edge will affect the accuracy of the sledged plaster and a perfectly true straight run will be impossible. To produce sections which are not straight the running edge of the template

52 Turning plaster on a lathe. A template may be set in the tool-holder on the left of the lathe or the tool rest on the right can support a hand-held chisel.

may be moved against a profile shaped to give the desired effect. This technique can be extended to the use of profiles to give models which are not simple straight horizontal runs but vary in height along their length. The variations and permutations are infinite, and capable of great subtlety and richness, but the skill and care required at all stages of the making of such models becomes very demanding.

Turning small-radius models is best carried out on a lathe. Wood-turning lathes can be suitable provided that they can run at a speed low enough for the turning of plaster. The plaster is cast either on a spindle or on an open chuck, the latter having the advantage that there is no central core to get in the way if you deviate from your original design during the turning process. In either case the plaster is cast in the normal way, using a 'cottle' or retaining wall of vinyl or other plastic; as soon as it has started to set the cottling is removed and the spindle or chuck set upon the lathe. The actual turning is carried out either by a hand-held chisel or by means of a template held in a tool holder so that it can be tracked in against the plaster while the lathe is spinning.

Great care must be exercised when using a hand-held chisel. At all times you must use the proper tool rest and maintain a firm grip so that there is no chance of the tool being pulled out of your grasp. A chisel flying across a workshop may prove lethal, and is at least a very frightening experience. Also, the chisel should be properly made with a deep steel shaft fastened into a wooden handle. Woodworking chisels are most suitable but it may be necessary to modify the shape of such tools so that the working edge conforms to the type of turning you wish to carry out.

The horse

Larger radii should be turned by means of a horse, which is very like a sledge in that the plaster is shaped by a template. The difference is that the frame holding the template is fastened to an arm which revolves upon a spindle. A horse is normally used on a large plaster-topped table, the plaster having been coated with several applications of shellac so that plaster poured upon it will not stick. When the template has been prepared and fastened to the horse, plaster is mixed and poured upon the table in the appropriate area. As the horse is moved around the central pivot an arc of the desired size is produced with the section of the template.

The production of larger turned pieces, which cannot be turned upon a lathe because of the leverage exerted upon the chuck and do not lend themselves to being produced by means of a horse, is possible by using a turning box. This is an open-topped box with slots in the two shorter sides which serve to support the turning axle. One of the longer sides is cut down to allow for the placing of a template so that it lies upon the same plane as the axle. Templates may be made up in the manner already described but the spindle is removed to allow the first layer of plaster to be cast upon it. Subsequent layers are cast on to the axle while it is installed in the box so that by turning the axle by means of a handle fixed to one end the plaster will be shaped to produce a column with a profile determined by the template.

Rectangular blocks cannot be cast with accuracy and will normally need to be trued up after casting. Most of the excess may be cut away by means of a bandsaw or, if this is not available, by planing with a pierced file such as a 'Surform'. To ensure that you do not remove too much, the plaster should be marked up with an indelible pencil; this allows for the plaster to be damped if necessary at a later stage without the markings being lost. Even greater accuracy can be achieved by soaking the plaster and grinding it on a ground-glass plate. The glass must be lubricated with a little water to prevent the plaster from binding to the glass and to permit the material ground off to be pushed away from the model.

53 Plaster of Paris models, about 140 cm (4 ft 6 in.) high, used in the restoration of the Natural History Museum, London (see p. 13). From these models, moulds were made.

54 Grinding a plaster block model on a sheet of glass with a sand-blasted surface. The model must be thoroughly soaked to keep it from sticking to the glass, which is also liberally sprinkled with water.

Plaster blanks for tiles may be produced by the method described above and any relief modelling introduced by drawing the design on to the plaster surface so that it may be modelled by hand, using rifflers and steel modelling tools. Indeed all the sections fabricated by any of the previously described methods may be given modelled details by drawing and remodelling by hand on any of the faces. Commercial modelling tools provide the basic shapes and serrations but if this type of modelling is to be an important feature of your work it is most unlikely that they will be sufficient in themselves. Special shapes can be made from metal sheets and again zinc litho plates are ideal. Such tools are more like small templates than the cast or forged modelling tools but they can be made if you have the facilities. Cutting tools will be necessary for some types of modelling and dentists' tools as well as surgical scalpels may be seen among the equipment which anyone professionally involved with modelling will possess.

CASTING TECHNIQUES

From time to time it may be necessary to cast either from a previously made plaster model or from a ceramic piece (this is certainly necessary when restoration work is undertaken). Traditionally plaster would be used but this entails cottling the area to be cast as the plaster takes quite a long time to set. In many cases rubber is a more suitable material, particularly when the piece to be cast includes undercut areas. Hot rubber (melted to a liquid and applied hot to produce an accurate cast when cold) may be used in some cases and when the rubber cast is no longer required it can be reheated and used again. This type of rubber must be handled very carefully and you should protect hands and eyes so that any accidental splashes do not cause severe burns. Cold casting rubbers are now available and although they cannot be reused they provide a most convenient material for safe and accurate casting. In both cases when rubber has set it is simply peeled off the surface and filled with plaster to produce a facsimile.

In restoration work the original piece will probably have been damaged and the purpose of the plaster model will be to repair the damage evident in the plaster and then to make a mould of the repaired model so that a new ceramic piece may be produced. The repair is made by casting plaster against the damaged area. The plaster model must be damp if the fresh plaster is to hold fast; alternatively, a metal pin can be driven into the model and held with a suitable adhesive. The fresh plaster need not be permanently fixed to the model but can be registered by means of its shape and the pin. The fresh plaster, in either case, may be modelled to the right shape and then permanently fixed to the model with an adhesive, after which the repaired model is used to produce a mould.

Whenever plaster is to be used as a model to form a plaster mould it is essential that the surface of the model be correctly treated. To the novice it seems inevitable that whenever you want plaster to hold a new layer it rejects it, and whenever you want it to resist a fresh mix it sticks. So the treatment of the plaster model must be carried out with great care to ensure that many hours of work are not lost at the final stage by the model being locked into the plaster of the mould. Traditional treatments for rendering plaster impervious, so that it can be cast against without sticking, are coating the mould with shellac or soft soap. Soft soap should be heated, put through a 100-mesh sieve to remove any lumps, and smoothly applied with a soft sponge to a well-soaked model. Several applications may be necessary, the excess being wiped off after each application with a damp sponge. The final result should be a plaster surface upon which water will lie only as droplets, and show no signs of penetrating the surface.

With larger models, shellac is used; this is heated and applied, usually, with a soft paintbrush of a suitable size. Again, several applications may be necessary and the result-ant plaster surface will be yellowish-brown. More recently, epoxy resins have been used to treat plaster, and these have the advantage of strengthening the plaster as well as render-ing the surface impervious. The greater cost may inhibit the widespread use of this type of resin but it still has a place in the treatment of plaster models.

When the plaster has been treated the next stage is to decide upon the type of mould best suited to the method of producing the clay cast. Moulds for slip casting should be uniform in section throughout so that the cast layer of clay will be even, although this is less important with large, heavily grogged casts. Press moulds must be strong enough to withstand the pressure of the clay being pressed into them. The shape of the model will determine how many divisions there will be in the mould and where these divisions will occur. Obviously the mould must divide, so that it will come easily from the model and the clay may be cast and removed without damage. Round models can usually be cast in two pieces.

Complex models may be made by combining turned shapes with sledged or hand-modelled pieces before treating the surface and making the moulds, but it may be more convenient in some cases to treat each part separately and make several moulds, joining the clay pressings before firing, or even to fire them separately and join them together when the parts are installed.

9 Large-scale ceramics

Any attempt to describe all the possible methods of making large-scale ceramics would require several chapters, and probably need to be amended every few years. However, I shall describe in this section some of the more important industrial and craft methods which are applicable to a studio situation.

The simple hand processes of building up clay forms, i.e. coiling and slabbing, are all capable of extension into the making of large-scale ceramics, although several modifications and refinements may be necessary.

Some simple but important points should be noted. The amount of clay required is always far greater than you expect, when building a large-scale object. As the scale of the object increases so do many of the tools; cutting wires may be feet long rather than inches, and modelling tools and scrapers may be of the type used in the building trade, such as trowels, rather than those of the sculptor and potter. All these factors demand far greater physical effort than is necessary for the production of pottery. Also, lifting several large sections in or out of the kiln, even with the assistance of several people, may be more exhausting than you expect. It is therefore unwise to embark on any venture of this kind without adequate preparation.

It is essential that you have a realistic drawing to work from, which you can demonstrate to others who will inevitably be involved at some stage, showing the precise nature of your object. This preparatory drawing becomes even more significant when you consider that making large objects of any material can create storage and transport problems involving considerable expense and complexity.

If possible, first find a site which can accommodate your sculpture. Acquire the permission of the owner to place your work on the chosen site and, if possible, sell the object to him. He or she is unlikely to give any commitment, even to accepting a free gift or temporary loan, unless you can provide some idea of what the final appearance is likely to be. So your drawing should reveal, to you and to others who are less accustomed to looking at drawings, what the basic dimensions will be. The colour, or colours, quality of surface, structural and physical stability and weight, as well as the over-all content in terms of symbol or meaning, should be clearly demonstrated either in the drawing or in calculations which accompany it. There is nothing worse than making a piece weighing more than a ton and finding

55 Coade stone statue of George III in Weymouth. The polychromatic surface is achieved by means of paint.

56 Sarcophagus of Seianti Thanunia Tlesnasa, c. 150 B C. This Etruscan clay sarcophagus stands 122 cm (4 ft) high. British Museum.

that it is unacceptable to the owner of your site: it is not like a pot which, if unsold, can be returned to the shelf.

The stages of manufacturing the object should be discussed with other ceramists, so that you might arrive at the best technique, or combination of techniques, for your design. Not only the outer surface, but any necessary interior structures, need to be identified and designed to give maximum strength with minimum weight, and this will lead to discussions about possible clays to use.

CLAYS FOR LARGE STRUCTURES

In a studio situation it is best to use a clay that you know well. A clay which is new to you should be tested by making several pots or sculptures in a variety of techniques, so that you can become attuned to what that particular clay will and will not do. The problems of large clay structures are those of normal clay objects magnified many times and a mistake, an error of judgement, or lack of forethought may reduce months of work to a pile of hardcore, which you then have to remove.

One of the first points to consider is concerned with structure: the walls of the object must be strong enough to withstand its compressive weight. This means that they may have to be several inches thick. Also, if the object is enclosed the movement of air is restricted if not actually prevented, and drying the clay throughout its mass will be very difficult. In this case, the tendency will be to select a clay which has a high proportion of coarse material, such as a fireclay or grogged body, although this will inevitably mean a coarse-textured surface which may be unacceptable in the finished object. Problems of drying and the accompanying shrinkage will require considerable thought and discussion when selecting the clay.

If the work is to stand outside it will be necessary to ensure that the porosity and permeability of the clay are reduced to the point where the finished piece will not 'frost'. This may be achieved if the composition of the clay body is such that it vitrifies without deformation at an acceptable temperature. Alternatively, the clay may be treated after firing to seal all the pores in the clay wall so that while the physical structure of the clay is not changed it is made waterproof. This treatment can be expensive and time consuming, but may be unavoidable as the alternative is to fire the piece to maximum vitrification, with all the accompanying risks of large or small disasters: if the piece is fired to such a high temperature it may split, deform or collapse.

In dealing with these problems it is interesting, if not essential, to see how they are resolved in industrial terms.

Among the most familiar objects in the countryside are the insulators, supporting high-tension cables from the arms of the metal pylons. Inside a power station even larger insulators are required, which may be twenty feet (6 m) or more high. The body of the insulators is a low-fired (1180° C, 2150° F) porcelain, fine-grained and dense when fired; they are covered with glaze and are absolutely non-porous.

Insulator porcelain – recipe

Ball clay	20–40%
China clay	17–35%
Quartz	20–40%
Potash felspar	20–35%

Such a body, having the desired characteristics in its finished state, would be very difficult to work in a studio. In a factory the pieces are made over a period of days only, and would be dried in standard conditions, with both temperature and humidity carefully controlled, until the piece is ready for a long and rigorously regulated firing cycle. Only if you can simulate these conditions (which can be done without a complicated technological back-up, but with extreme care) should you embark on a large-scale construction with a porcelain-type body.

The more likely solution, a coarse-textured clay, may provide the acceptable drying properties but will be more difficult to model if the particle size is greater than the detail essential to the character of the piece. In this situation it may be necessary to construct the object from finer clay in several separate pieces, which are joined together after being fired. By this means the wall section may be reduced to well within the easy drying limits of the clay, and when fired the lower section should have sufficient compressive strength to withstand the weight of the upper pieces when installed. The precise locations of such junctions must be carefully designed to afford minimum interference with the formal composition of the object, or else integrated into the surface characteristics. No flowing movements around or down the form should be marred by unsightly and ill-considered joins. In particular, the edges of the pieces to be joined should be an accurate fit across the mating surfaces, and in over-all shape. This may be difficult to achieve if there is any distortion during the firing process.

In order to ensure an accurate fit you should make a paper pattern of the top of the lower section at the point where the next section will mate when fired. To do this, take a piece of paper larger than the largest diameter and place it across the top of the constructed section. Then, with a needle, pierce holes in the paper at intervals of 3 mm, so that the needle passes through the paper and against the outer wall of the lower section. A similar set of punctures marks the inner wall. It may be necessary to fasten the pattern with large-headed pins pushed through the paper and into the top of the clay wall, to prevent it from being disturbed during the piercing process. Absolute accuracy is essential if the pattern is to be of real value. When the

57 Japanese Haniwa tomb figurine in clay of a warrior in armour of bamboo bound with leather; about third century A D.

inner and outer locations of the lower section walls have been marked, the pattern is removed, placed over a clean piece of paper of similar size and fastened with adhesive tape. The pattern is then pounced (see p. 128) through on to the clean paper. When the pattern is removed you can join the dots together with a paint or pen marker, chosen to contrast with the colour of the clay being used for building up the form. This contrast is essential if the accidental marks made during building are not to be confused with the lines of the pattern. The pierced pattern should be preserved, so that it may be used to produce another pattern if the first is damaged or lost.

Two points should be remembered if a paper pattern is necessary. One: make the pattern while the clay is still damp, so that you are measuring the plastic size and not the dry size, which will be smaller than that which you need. Two: do not make any modification to the lower section if at all possible. If such a modification is unavoidable be sure to mark it on the pattern and adjust the upper section accordingly.

The making of patterns is only necessary if the physical strength of the sections will not allow you to construct the object in one piece. If you need to cut the finished object in order to place it in the kiln the same guide lines apply about locating the division sensitively, but the cuts may be made with a knife or long cutting wire. In the case of the latter a sawing action will produce a cut in quite firm clay, but it will be impossible to cut dry clay without severely damaging the surface of the object.

There are clays available commercially which display remarkable dry and fired strength considering the thickness of the final section. One such clay is 'Treforest "T" Material', sold by Morganite Crucible Company. The texture is quite coarse but it will model to a smooth finish if necessary. The dry strength is important as it enables the relatively thin-walled (say between 12 and 25 mm) object to be carried and placed carefully in the kiln without damage. Although such clays are considered to be expensive when compared with other clays, this is no real disadvantage considering the advantages afforded by the physical properties, which greatly increase the chances of success.

Several experimental approaches may be attempted if you wish to formulate your own clay body. The ideal would be to have a fine-grained clay, with great dry strength and fired strength. The fired strength should be not only compressive but tensile as well. The object must be moved from the studio to its final site, and lack of tensile strength may lead to breakage of the object if it is moved from an upright position to lean at a slight angle. The clay should therefore have some proportion of fused material in its fired state, and felspar and borax may provide this. The dry strength may be achieved by the addition of organic binders such as polyvinyl alcohol and cellulose derivatives.

Some commercial manufacturers use sanitary china or electrical porcelain as the grog in hand building bodies as they will open the body during drying but will fuse into the rest of the clay if fired beyond 1200° C and no evidence of their presence may be found.

Before a large object is constructed you must consider how it will be moved into the kiln. The simple solution is to use a kiln with a trolley hearth, so that the object may be built on the hearth and rolled into the kiln for final drying and firing when the building process is complete. If such a kiln is not available, or the hearth may not be put out of use for the length of time it will take to construct and complete the object, then the building must be carried out on some other surface which can be brought into line with the kiln when the piece is ready for firing. If access to the firing chamber is possible only through the door on to a fixed hearth, ingenuity will be required to devise a method of transferring the dry piece to the kiln hearth.

One method sometimes favoured is to construct a trolley, either on wheels or rollers, from suitably strong timbers, or bricks, so that the structure and wheels will support the total weight during fabrication. This trolley should have a top surface which is exactly the same height as that of the kiln hearth. If the kiln is of the moving hearth type the top of the constructed trolley is covered with coarse sand or grog, under a thick layer of polythene sheeting, and kiln shelves placed on top to provide a building base for the object if required. On this the piece is built, with sand placed between the clay and the kiln shelves to avoid sticking, and taking care that the over-all height of the work does not exceed the distance between the hearth and the top of the firing chamber. It is possible to use dimensions which will permit the piece of work to fit the kiln chamber only when the shrinkage is complete. This approach is full of pitfalls to the beginner, and if the size of the piece is very close to the inner dimensions of the kiln, placing the piece in the kiln will require the utmost care.

Once building has been completed, and drying has reached an advanced stage, the trolley is moved so that it lies adjacent to the hearth, and by carefully pushing the object at its base while simultaneously pulling gently on polythene in the same direction, the piece may be made to slide on to the hearth, which will have been previously covered in a similar layer of sand or grog. Once the piece is accurately located on the hearth it may be moved into position within the firing chamber. During the firing process the polythene burns away and if the kiln is electrically fired the chamber should be ventilated while the polythene burns. This will avoid the danger of harmful gases building up within the kiln, and any reducing effects these may have upon the elements.

Moving large pieces in this way can be difficult and dangerous. The piece may be damaged or destroyed al-

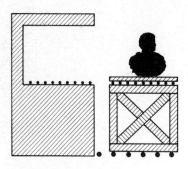

58 Moving a large sculpted piece into a fixed-hearth kiln by means of a specially constructed trolley and rollers.

together, and the lifting and sliding action involved may require the energy of several people.

If the kiln to be used is a fixed-hearth type the problems of moving the finished piece will be even greater. A similar trolley must be constructed, but instead of having sand and polythene lying beneath the kiln shelves there should be fireclay rollers or spheres. Porcelain balls, such as those used in a ball mill, may be used, but one disadvantage of using spheres as the wheels beneath the slab on which the object is built is that their ability to move in any direction may create a surface which is unstable during the building operation. Provided that the trolley can be accurately lined up with the kiln hearth, fired clay rollers provide the most suitable means of moving the base shelves into the kiln. It will be necessary to have some rollers placed in the kiln, so that the shelves bearing the finished object may be pushed forward without tilting and becoming unstable. These rollers should be placed at regular intervals across the kiln hearth before moving the finished piece into the kiln, as it may be difficult, if not impossible, to place more in position once the base shelf has been pushed part-way home. Clay rollers may be made by extruding through a suitable dye fitted to a wad box or pugmill. Tubular kiln props are suitable if there are enough to carry the compressive weight of the base shelves and finished object without being crushed. Once the piece is fired it may be removed from the kiln by reversing the procedure devised for placing it in the fixed hearth.

Removing heavy fired clay objects from trolleys may require a block and tackle or hoist, if you have such facilities available. If not, the piece must be moved on to a brick or wooden structure, built up in layers; then if the object is tilted a layer may be removed from the front, the object rested on the front stack, and a layer removed from the back, and so on. It is then transported either to the prepared site, or on to a small trolley to be moved out of the kiln area. Problems of moving large and heavy objects such as insulators are solved in industry by using fireclay palettes which are moved with great cate on fork-lift trucks. The kilns are intermittent electric or gas-fired, and large enough to allow the fork-lift truck to enter the chamber.

METHODS OF FABRICATION

The simplest method of building up a large ceramic form is a scaled-up version of coiling, except that the coils may take the form of large strips. If the piece is to be built directly on to kiln shelves, these should be coated with a layer of sand to allow the clay to contract and move on the shelves during drying and firing, without sticking or cracking. The coils may be 7.5 to 10 cm thick provided that the clay can be dried either because it has a coarse texture, or because the form is relatively open in design, and allows the air to

move easily across the surfaces. The clay should be fairly
soft, and joins between the coils should be removed by
modelling and smoothing with a flexible metal scraper.
The clay form may have to be left to dry from time to time,
before adding more coils and increasing the weight which
the lower section has to support. During this drying stage
the upper coils must be kept damp by covering with either a
damp cloth or polythene sheeting; this will ensure a strong
union between this layer and subsequent coils.

When the building process is complete the detailed
modelling may be carried out, which means that the complete form must remain in a plastic or semi-plastic state.
It is very easy to allow the base to become too dry, and
attempts to moisten the surface will probably produce
troublesome surface cracks. These are caused by the outer
surface being expanded by the addition of water, which
seeps between the clay particles. If this surface is then
modelled it will contract upon drying, and no longer fit
the inner layers of clay, and will either shell off or crack
apart. When the modelling is completed the piece should
be set aside to dry very slowly and evenly.

Slabbing may be used for appropriate forms but this will
probably have to be built in sections as the weight of very
large slabs will make them difficult to handle without their
breaking; the tensile strength of unfired clays is usually poor.
The slabs may be made by laying up either on a blockboard
(which, as it tends to warp and split apart if water penetrates
the laminated layers, must be protected from water impregnation by polythene sheeting), or on marine ply, which is
much more expensive. To determine accurately the thickness of the clay it will be necessary to fasten metal or wooden
battens on to the board, corresponding in height to the
thickness of the required slab. If the slab is to be thinner at
the top than at the base, so that it may more easily support
its own weight when upright, the battens should be similarly tapered by sawing or planing. When the battens have
been fixed into place with countersunk screws, the recess is
filled with a suitable clay. This should be slammed into
place, paying particular attention to the edges and corners
so that the slab will be an accurate reproduction of the
shape constructed on the board. Beaters, covered with
string or hessian to prevent them from sticking, will help
to drive the clay home, and if this beating is carried out
across the whole surface it will increase the density and the
strength of the slab. When the board is covered, a long
cutting wire should be drawn through the length of the
slab. This will probably require two people, one at either
side, and the ends of the wire should be pulled downwards
so that the wire runs along the edge of the batten, producing
a slab of even thickness. Any low points will be revealed by
this cutting; these should be filled and the cutting action
repeated until the slab is quite even.

Several slabs may be made in this way and, if space

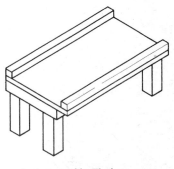

59 Laying-up table. The batons are
removable and the top surface
overhangs the table proper so that it
may be removed and the clay tipped
out on to the drying surface.

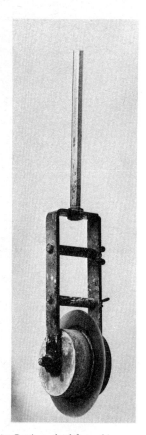

allows, they should be cut from the board (by removing the battens and drawing a wire between clay and board) and turned over on to a suitable plaster surface, thus reducing the risk of warpage due to the top surface shrinking faster than the lower.

While you are turning the clay over there is a tendency for it to flop from the board once it has been lifted to a horizontal plane. You can prevent this by running your finger round the edge of the clay when it has been cut from the board. This seals the edge of the clay against the edge of the board, which will be enough to hold the clay in place while it is being moved. Once it is against the drying bed the clay can be released by giving the laying-up board a sharp jerk. If necessary, the clay seal may be cut through with a knife or needle before shaking it free, thus ensuring that the clay will release first time.

When cut to shape and to a convenient size, the form can be fabricated in the usual way, using cross-hatch and slurry techniques to join the pieces together. Both the inner and outer surfaces of the joins should be filled with clay, beaten home, and modelled or planed to a smooth surface.

60 Cutting wheel for making a vertical cut on large clay slabs. These, when partly dried, are laid on plaster slabs for cutting. The distance between the cutting edge and the roller is the thickness of the slab.

61 Casting tiles or blocks with fireclay. The slip is held in the high-level mobile tank, and is fed into the mould through a funnel and a hollow plaster block. The second block acts as a riser: when the mould is full the two hollow blocks are topped up with slip to keep the contents of the mould under pressure. This ensures that no air can enter and the blocks cast solid.

If the form is to have a flat top it will be necessary to provide a rib-vaulted supporting structure, as large slabs may collapse or crack during drying and firing when supported only at the edges. Alternatively, the top may be made and fired separately and fixed into position when the form is complete. In this case a pattern should be made of the top of the upright form in order to ensure even a reasonable, let alone accurate, fit. In the chemical stoneware industry it is not unusual to construct rectangular acid tanks in this way, using hand methods only.

It may be desirable to fabricate a large object, either totally or partially, by a moulding process. In this case a slip-casting mould presents complex problems and involves very large quantities of plaster and slip. Press moulds are more appropriate for large-scale work. These moulds are usually made in pieces by casting from the relevant model; the mould, which may be less than 2.5 cm thick, is then strengthened by bonding laths and scrim on to the outer surface. The scrim should be cut to suitable lengths and dipped in a fresh mix of plaster, so that it becomes thoroughly impregnated. It is then laid over the plaster which is being cast on to the model, and used to secure the wooden laths in place.

It is not necessary to produce a mould with an even build-up of plaster as the porosity of the surface plays only a minor role in shaping the correct clay form. Once the mould is made, the clay is pressed into it and removed as soon as it is strong enough to support its own weight without being deformed.

During the pressing of clay in plaster moulds, the surface which is in contact with the plaster is subjected to immediate drying, called 'surface drying'. This serves to stiffen the clay so that it will retain its shape in the mould without the necessity of building an internal structure for small undercut areas. In the hand pressing of chemical stoneware, which is usually concerned with making relatively smooth forms, templates are used to smooth the interior of the clay; this ensures an even section, and may be used to produce a ribbed interior, which is structurally stronger than a smooth surface of the same thickness. All this is done by hand, and requires real skill but allows for great flexibility of design.

If left too long, the clay may lock into any undercuts and be impossible to remove. Also, it is very likely that clay pieces made in large moulds will be damaged to some extent during the removal from the mould. This will almost certainly be the case with moulds made from natural objects. When the pieces have been moulded they may be assembled and joined in the same way as flat clay slabs, and joins or damaged areas modelled out. With this in mind it is sometimes easier to aim to produce the general characteristics of the desired form when making the mould, and to model details and the surface when the clay form has been assembled.

62 Polystyrene model of 'Gugis' the gorilla, made for John Aspinall's zoo near Canterbury, Kent.

63 Pressing clay into the inverted mould of the gorilla.

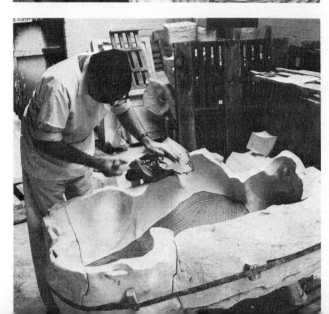

64 Second layer of the mould assembled: pressing continues.

82

65 Third layer partly assembled and in the process of being pressed.

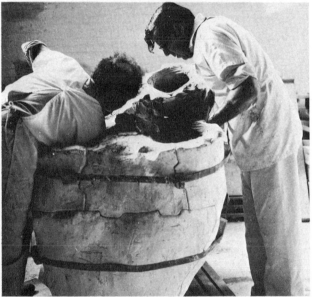

66 The complete mould assembled: pressing is finished off through the feet and legs.

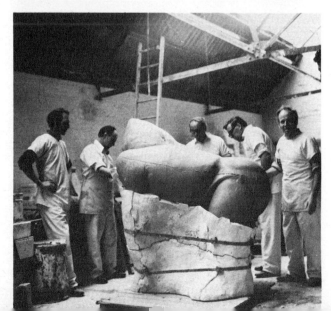

67 The mould turned the right way up and partly disassembled.

68 The mould removed and detailed modelling in progress. As the clay is still not strong enough to stand by itself, the pressing is supported.

69 Detail modelling is finished and the surface is coloured with oxides to produce the desired effect. The model is set on a large clay shelf, under which are pads of unfired clay, then a layer of brick, then rollers to allow the piece to contract in all directions, and finally a layer of brick.

Where a symmetrical form is wanted, turning is the recommended method. In the manufacture of insulators, for instance, large de-aired slugs of clay are extruded and stood on end to dry until they are firm to the touch – a matter of judgement on the part of the turner. When he considers it to be in a suitable state the clay is moved on to a jig so that it sits upon a turning wheel. The jig includes

a template and a ring turning tool. The clay, spinning on
the wheel and at the same time being trimmed with the
turning tool, will be turned into the required profile. In
the early stages of turning, the tool guide touches the
template only occasionally, but towards the end of the
process the tool is running hard against the profile, thus
producing the exact shape required. The insulator is finished
by using a smoothing tool and sponge, then cut from the
wheel with great accuracy.

Very large insulator sheds are made upside down by
jollying large slugs into plaster moulds and turning the
ribs with a template. This process is a combination of
throwing and jollying, and requires great skill and accuracy.
These sheds, when leather-hard, are stacked one upon the
other in a turning jig and trued up by hand. The slip for
joining the sheds is coloured to reveal the joining zone.

When complete, each insulator is placed upon an unfired
clay bat which remains beneath it until it has been fired.
This bat shrinks at the same rate as the insulator and acts
as a buffer between the insulator and the kiln bat which
supports the assembly during firing. Very large insulators,
6 metres high or more, are made in several sections and
assembled in the kiln. To bond the separate shed stacks,
they are fired together with a glaze very similar to the clay
body, and the two are indistinguishable after firing.

70 Turning an insulator from a
hollow extrusion. The turning tool
is held on the arm in the foreground
and is guided by another arm at right
angles to it which runs against a
template. The turning tool and guide
are raised and lowered by means of
the large wheel on the right.

85

71 Turning an insulator: view from the back, showing the template in the foreground.

72 Horizontal turning on a small lathe. The guide is above the lathe and the turning tool is running against the clay.

73 Jollying an insulator shed. The
clay is set in the mould by hand, and
raised and turned into shape using
the template, while the clay is kept
damp by the operator.

74 Insulators ready for firing. Some
in the foreground are set in saggars.
The one in the middle was too tall
to be fired in one piece, so after
firing, and consequent shrinking,
both parts were set together with
glaze between them and refired.

INTERNAL STRUCTURES

In some cases consideration should be given to the desirability of incorporating an internal structure, either as a permanent part of the object or to be removed once the form is strong enough to support itself (which may only be after the firing cycle is complete). In some cases a system of buttressing and vaulting may be sufficient to support overhanging or weaker parts, such as joins between prefabricated pieces. In others, it may be necessary to construct a self-supporting skeleton of clay to tie the surfaces together, and provide support where needed. If possible these buttresses and internal structures should be hollowed out or shaped, so that they will accommodate the compressive weight without undue increase in the total weight of the object. This may sometimes produce something more akin to aircraft construction than clay work. If the internal structure is intended for removal at some stage it is better to make it in several pieces to prevent it from being trapped within the form by some unexpected change in shape either of the structure itself or of the outer form.

DRYING LARGE CERAMICS

The ideal system of drying is to have an area where the humidity is controlled, and simply to reduce the humidity very slowly until the clay form has dried. Even this might not remove all the water contained in the clay within a reasonable length of time. In industry, infra-red driers are sometimes used, consisting of a number of infra-red wall heaters in a room isolated from the rest of the factory. The rays given off from these heaters are harmful to human tissue if exposed for any length of time, but they have the effect of penetrating the clay and heating the water within it, thus driving it out rather than relying solely on the capillary action of normal drying. Less effective, but with a long tradition in smaller specialist factories, is the simple expedient of hanging one or more light bulbs within enclosed forms. These give off a small amount of heat when switched on, and the air currents produced by this method draw out the water from the inner surface and carry it into the atmosphere. Other factories use under-floor heating (usually from pipes carrying hot water or steam) which can be increased or decreased according to the progress of the clay being dried; the drier the clay becomes the warmer the floor, until drying is almost complete. Others use drying tunnels which gradually heat the clay as it passes through the tunnel.

Drying can only be completed in the kiln as part of a slow and careful firing of the form, which may take many days to raise the temperature to the maturing point of the clay.

75 Drying rectangular stoneware tanks: the arrangement of blocks of unfired clay, fire-bricks and fired clay rollers allows the tanks to contract in all directions.

During the drying of the clay object it will be necessary to pay particular attention to the sharp edges, or modelling, on the form. If the piece has to be handled during this drying process there is every likelihood that such crisp details may be chipped or broken, if subjected to careless handling. When this occurs it is extremely difficult to repair the damage with plastic clay as the bond between the drying clay and the plastic clay is unlikely to be strong enough to withstand the shrinkage differential. Experienced builders of large-scale ceramics sometimes make allowance for warping during drying or firing. Therefore it may be necessary to exaggerate a convex or concave curve in the form so that it may move into the designed shape by the action of the shrinkage or partial collapse of the clay during drying and firing.

76 Dutch tile panel 181 cm high by
144 cm wide. Dated 1737, it shows
the manufacturing processes used in
the Bolsward tile and creamware
factory, Amsterdam. At the top are
the arms of the factory's founders.

10　Tile-making

There are many ways of making tiles. They may be made from plastic clay, slip or dry clay and, in theory, be of any size. However, the larger the tile the more difficult it is to produce a finished object which is flat and dimensionally accurate. Large tiles may make the work of the designer easier but they create many problems during drying and firing. Traditionally, tiles were limited in size so that they could be handled easily and any slight malformation in one tile would not entail remaking the whole design.

CLAYS FOR TILE-MAKING

Before the industrialization of tile manufacture natural clays, occasionally with some other additions, were used for tile-making. Suitable deposits would be secondary clays with fine grain structure, to which could be added grog or sand. The process of making tiles is carried out with a good deal more care than brickmaking, and if the clay being used was washed into rivers from higher ground during the rainy season, as it was in some parts of Italy, the clay–water mixture would be run off into settling beds by diverting the flow of the river. When these beds were full the water would be diverted to its original course, and the clay allowed to settle while the water evaporated or was drawn off through small channels. The clay might be allowed to remain in these beds for several years. This ageing increases the plasticity of the material and is known by several terms, such as 'weathering' or 'leaning'. Where no such natural agency as a convenient river is available suitable clay deposits may be found in the courses of dried-up rivers, or in the beds of lakes where the geography of the area has changed to leave a deposit of secondary clay without the water which, at one time, covered it. These deposits may include stones and other impurities which can be removed either by wedging or filtering the clay.

Industrial tiles are produced from plastic clay only if the method of manufacture is to be the extrusion method. Most factories have gone over to dust pressing and the body for this process is carefully controlled. It will most likely be a mixture of ball clay, china clay, flint and felspar, formulated to produce a white-firing body giving satis-

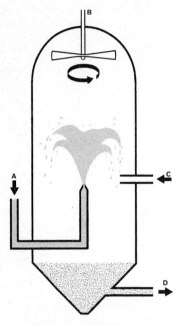

77 Spray drier. The slip enters
through A, emerges from the pipe
as a fountain, is atomized by the fan
B, and as the droplets fall they are
dried by the hot air entering through
C to fall as granules and be drawn
off through the pipe D. In some
designs the slip enters through the
top of the drier, directly into the fan.

factory mechanical strength when fired to 1050° C
(1920° F). These ingredients will be carefully prepared
by crushing or grinding them to produce a satisfactory
particle size. China clay is won from the site of its decomposition by washing it away with high-pressure water
jets. As a primary clay it is white-firing and relatively
refractory. The ball clay is a white-firing clay, and gives
plasticity to the body as it is a highly plastic secondary clay.
The flint is produced from flint pebbles which are calcined
and crushed. These, together with the mineral felspar, are
blended in a large mixing tub, known as an 'ark', in proportions determined by the recipe for the finished body. They
are mixed with water to produce a slip and blunged, or
mixed, for several hours to produce a well-blended suspension. After blunging, the slip is pumped to a filter press,
where the water is extracted, leaving plastic pancakes of
clay known as 'cakes'. These cakes are then dried artificially,
crushed, and passed to the presses. The water content for
press tiles is carefully controlled and may be between 10
and 20 per cent by weight. More recently, air drying
techniques with spray driers, in common use in the food
industry to produce dehydrated foods, have speeded up
the conversion of slip to dust particles.

Flow characteristics are an important factor in bodies,
particularly those which are to be used for dry and semi-
dry pressing. A varied particle size will produce an acceptable compromise between the small particles for close
packing (high density and detail reproduction from either
the die or mould) and large particles for easy die filling.
In order to obtain the necessary flow properties in fine-
grain material, the powdered clay must be agitated, which
is done commercially by mechanical vibrators.

In small-scale production the condition and working
properties of the clay body can easily be monitored, and
unsatisfactory qualities quickly modified. If the body will
not produce details, then a higher proportion of fine material
should be present. If it tends not to move easily in the mould
or die, an increase in the proportion of coarser material may
encourage the body to flow when pressed. These qualities
are a major factor in automated production where material
is moved by conveyer, or piped under pressure, and in the
tile industry where automatic die-filling is necessary.

Lubricants and binders are used commercially for pressing
bodies. Two parts of paraffin to one part of a water-soluble
oil, such as a machining coolant, and ten parts of water
have been found to improve the flow characteristics of the
body and the strength of the pressed article without
increasing the water content to a point where drying
presents problems.

In other cases oil alone can be used. Paraffin wax (up to
10 per cent), either as flakes melted into the body or
emulsified, is also used to bind dry pressing material, as is
polyvinyl acetate and cellulose derivatives.

Top: the Auckland Arms, a 19th-century Portsmouth public house, now demolished. Polychrome glazed brick and faience. *Above*: terrace floor, Osborne House. Encaustic tiles by Herbert Minton, *c.* 1850.

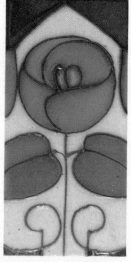

Top row, left to right: tube-lined tile, late 19th century; relief tile by Minton, 19th century, dust-pressed with polychrome glazes; encaustic tile, plastic-pressed, 19th century; under-glaze transfer tile, early 20th century.

Bottom row, left to right: under-glaze direct-printed tile, dust-pressed; under-glaze copperplate transfer hand-coloured tile, early 20th century; dust-pressed relief tile, green glaze, early 20th century; relief tile, dust-pressed in imitation of tube lining, glazed with two colours.

Right: trial glazes for reactive colours. Four variations covered with one glaze (H. & R. Johnson, 1977).

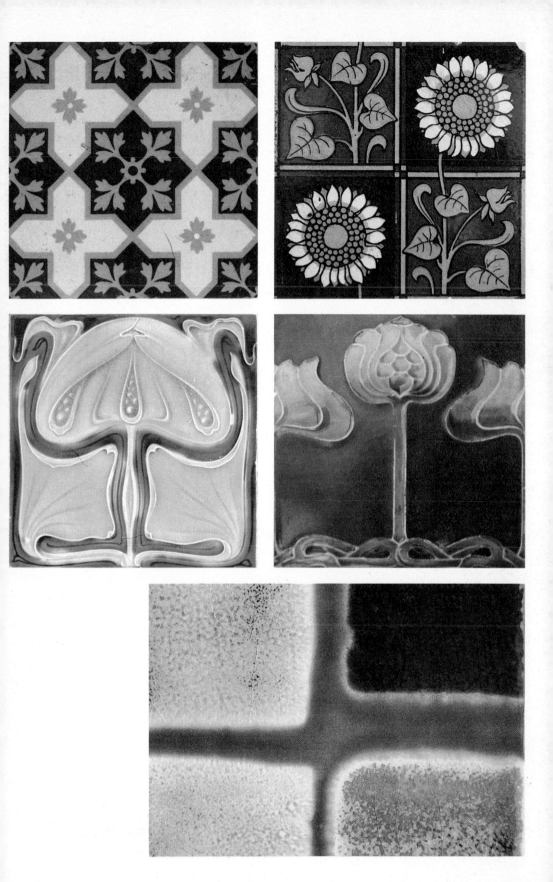

Sun Wall by Miró and Artigas. UNESCO Building, Paris, 1958.

Terrace wall designed by Gaudí, Parc Guell, Barcelona. Broken industrial tiles, with some made to specification. Early 20th century.

Tiles which are up to 15 cm square would normally be made about 9 mm thick. Tiles larger than this will need to be proportionately thicker so that a 60 × 30 cm tile might be 2.5 cm thick. A large tile with a thin cross-section would be difficult to transport to the kiln as it would be unable to support its own weight when carried horizontally. Similarly the stresses during drying and firing may be such that they build up forces across the face of the tile which cause it to crack if there is not enough material there to withstand these pressures.

MAKING TILES FROM PLASTIC CLAY

Clays suitable for making tiles from plastic clay should contain some coarse material to produce capillaries large enough to permit the water to evaporate from the clay with relative ease. The clay should be plastic enough for it to retain its shape having been formed, but a clay which is too plastic will shrink excessively and warp. Tiles with a high shrinkage are difficult to dry evenly and any surface which dries first will shrink, thus creating a curved tile with the drier side producing the inside curvature. The colour of the clay should be appropriate for the decorative process involved, but this is a secondary consideration as the surface of the tile may be coated with an engobe of a suitable colour, if the colour of the body is a disadvantage. Earthenware and stoneware clays normally used for throwing may be used for making small tiles – up to 15 cm square – with the addition of fine grog. Tiles larger than this should be made from a clay which includes in its composition sand and/or grog with a relatively large particle size.

The simplest method for making tiles which might be used for glaze trials is to roll out the clay on a hessian-covered board to which have been fixed two guides of identical thickness so that the rolling pin may be used to press the clay down as far as the guides. The rolling motion of the pin will squeeze and compress the clay to an even thickness but irregular shape. The clay should be left to dry to a leather-hard state, when it may be cut to the appropriate sizes and shapes. This drying should take place slowly to avoid warpage. As soon as the clay has been cut to size the individual tiles should be lifted and turned face downwards so that the back surface is allowed to dry. When both surfaces are of equal dryness the tiles should be stood on edge to allow an even flow of air over the surfaces. When the tiles are thoroughly dry they may be placed in the kiln and fired.

If more than half a dozen tiles are required, rather than roll out large areas of clay, it is easier to knock up a rectangular lump of clay and to cut this into convenient layers, slightly thicker than the guides will produce. This slicing may be carried out on a wooden board and the thicknesses

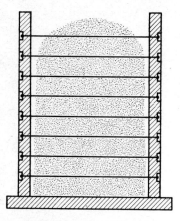

78 Tile-cutting frame. The clay is set on the base and the frame, with cutting wires in the measured slots, is drawn across the board so that the wires cut through the block of clay. In some designs the cutting frame may be fixed and the clay pushed through it.

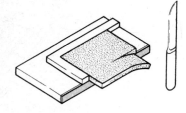

79 Tile trueing frame. The frame consists of a steel-edged board with a try square fixed on it.

of the slices can easily be regulated by using two pieces of wood about 20 cm long with a square section of about 2.5 cm. These measuring sticks should have slots cut into them at regular intervals. Slots should not run any farther than halfway through the thickness of the wood. They can easily be made by clamping the two pieces of wood together and sawing the slots with a tenon saw, having marked the position of the cuts with a pencil and ruler. When these guides have been made a strong piece of wire, such as piano wire, with toggles attached to the ends, is located in the bottom slot of each guide and pulled tight so that when presented to the face of the clay block, with the bottom of the guides resting on the board, a smooth even pull towards you will draw the wire through the clay to produce a fine cut parallel to the board. This process is repeated, with the wire resting in successive slots, to produce a block of clay cut into even slices. The top layer may be discarded and subsequent layers placed upon a laying-up board between the measuring sticks and rolled flat. The rolling action is important as it produces a more compact tile and the surface will consist of platelets of clay orientated with their faces uppermost.

When the clay is leather-hard it should be trimmed to size. If many tiles are to be produced it is worth while to construct a cutting board, which consists of a square piece of wood with a right-angled frame attached to it, and a metal edge attached to one of the open sides of the board. When an untrimmed, reasonably accurate, clay tile is placed against the frame a sharp knife may be drawn towards you, running against the metal edge of the board, and this action will remove any irregularities on the side of the tile. This process may be repeated on each side of the tile when the clay is in a leather-hard state, and will ensure that all tiles produced are dimensionally accurate.

Alternatively, templates may be made, either of wood or plaster. These should include some kind of knob or small handle on the back and are then presented face downwards on to the clay. A knife is run along the edges of the template cutting through the clay. The template may be removed without damaging the tile beneath by using the knob or handle on the back.

Tile frames

Tile frames may be made from wood, metal or plaster. They consist of the frame in which the clay is pressed, and the block of the same material cut so that it will slide through the frame. The frame is located on a suitable laying-up board, with canvas or polythene to ensure that the clay will not stick to the board. Clay is then pressed into the frame and if the frame is of metal a roller can be run across the clay. The pressure will squeeze the clay against the edge of the metal frame, simultaneously cutting it to size and

80 'Holloway Doors' by John Mason, 1962; ceramic, 213 cm × 163 cm (84 × 64 in.).

pressing it home. If the frame is made of wood it is easier to overfill the frame with clay, which is pressed well home and then trimmed to suitable thickness by drawing a cutting wire, or 'harp', across the surface of the frame to cut the clay to the desired thickness. The top layer may be removed and discarded, or used for subsequent tiles. When the clay has been allowed to dry until it is almost leather-hard, the matching block should be located on the clay tile and held in position while the frame is pulled upwards, leaving the clay upon the laying-up board and perfectly flat.

LARGE WALL PANELS

Where a large one-off panel is being made it is often desir-able to make this in one piece and cut it into sections for easy handling only when the design has been completed. The finished dimensions of the panel should be used to calculate the clay size, which may be 10 per cent or more larger than the finished piece. A board of suitable size should be placed across several benches or on the floor of the studio. To this board you attach long guides which will produce a clay slab of the desired thickness; these guides should be fastened to the board either from the front or from the back, using countersunk screws. This frame may be used to hold the polythene necessary to separate the clay from the laying-up board; alternatively, sand or grog should be sprinkled on the board within the frame to inhibit the sticking of the damp clay to the porous board. Chipboard may be used for the laying-up board, but this is rather porous and may warp if water penetrates the surface. It will certainly flex if it is lifted with the clay on it. This may result in damage to the clay once it has started to dry, so if it is likely to happen the board should be battened on the back to produce a more rigid structure. The clay is laid up in the normal way and pressed home, using wooden bats which have been covered in hessian or rope to prevent the wood sticking to the clay. The bats should be about 35 cm in length and easy to handle so that the swinging and hammering action will not require any more of your energy than is absolutely necessary.

When the board has been over-filled the clay can be trimmed to the correct thickness using a length of cutting wire with strong toggles securely attached. Trimming the clay with a cutting wire normally requires two people, one at either side of the board, holding the wire tight and pulling it down at the toggle ends, so that the wire runs across the wooden guides. Inevitably there will be some drag as the wire is pulled across the surface of the clay, so that when the top layer has been removed it will reveal a rippled surface rather than a flat one. Recutting the surface may improve it but it may be necessary to produce a perfectly flat surface with the aid of a large roller followed by a 'kidney'. The kidney, made of flexible steel or rubber, is drawn across the surface of the clay to smooth any irregularities.

When clay has been pressed into any frame, whether small or large, there may be small fissures left in the reverse side of the tile, which will only be revealed when it is taken from the frame, probably in a leather-hard state. These fissures must be filled before the tile is fired, as they are points of weakness and may produce cracks during drying and firing. You should always keep some clay which has been used for laying up the tile and allow it to dry at the same rate at which the tile is drying. This may then be used to repair any weak spots on the tile, should they be revealed at a later stage in the making.

Any decorative process which is carried out on a plastic or leather-hard tile may distort the shape. Whenever possible, this type of decoration should be completed before the tile is finally trimmed. If the tile has been made in a frame the decoration should be applied before it is removed from the frame. If the decoration is such that pressing the frame

81 Pugmill for the extrusion of twin tiles (the die has been removed). The mill is fed by conveyers which carry the clay to a hopper.

block upon the surface of the tile will destroy or damage it, the frame should be removed prior to decoration and the tile trimmed by hand afterwards. In all cases allowance should be made for the shrinkage during drying and firing. The dimensions of tile frames and moulds should be calculated to produce accurate clay size which will result in an accurate finished size.

Commercial frames and cutters are available, but most of these are suitable only for limited studio use and are rarely seen in workshops committed to commercial tile production.

EXTRUDED TILES

In recent years the technique of extruding clay, originally developed for brick production, has been modified for the production of extruded tiles. The clay is fed into a hopper and extruded through a de-airing pugmill. Instead of having a simple square-section die, as for the extrusion of bricks, the mouth of the pugmill supports a die plate so shaped that the extrusion takes the form of twin parallel slabs of clay joined together with dovetails cut partially through by vertical knives in the die. The extrusion feeds on to a conveyer monitored and controlled by special jigs which measure the appropriate length of tile and wire-cut the twin extrusions before passing them on to be stacked and set for drying. Such tiles are marketed under the name of 'twin tiles' and can be found in various sizes according to each manufacturer's specifications.

Obviously the clay must be de-aired to ensure a consistent extrusion. However, the passage of the clay through the augers in the pugmill tends to generate a laminated structure with an inbuilt 'memory' of the twisting to which the clay has been subjected as it is driven through the mill. Sometimes adjusting the distance between the end of the auger and the die plate will allow the clay to settle into a more amorphous mass before being shaped by the die. It is not unusual to find that the back of the die has several plates shaped to modify the rate of flow in various sections of the pugmill. These are adjustable to effect a re-orientation of the laminations of the clays to produce a straight extrusion.

To ensure an ideal extrusion the die must extrude the clay at the same rate at every part of the die. Where the die is producing a very narrow extrusion the rate of flow of the clay will be faster than that through a wide opening. The interrupter plates at the back of the die serve to modify and balance the flow of the clay so that as it comes from the various parts of the die the tendency to warp and twist is kept to a minimum. Without these plates there is a tendency for the extrusion, as it comes through the die plate, to twist in the direction in which the augers are turning. No matter how sophisticated the jig devised to receive the extrusion, clay with an incorrect orientation will produce tiles which will twist and warp during drying and firing.

As the tiles are extruded back to back they may travel on conveyers or carrying boards resting on the narrow edge of each tile. This allows for the maker's name to be impressed by an engraved roller, either before or after entering the cutting jig, and the two tiles are relatively safe to handle without distortion. The extruded pieces are quite strong and as the tiles are back to back the risk of their warping due to any accidental irregular drying is reduced.

When the tiles have been dried, either by atmospheric drying or through specially designed drying cabinets, they pass into the glazing area. This is a series of automatic airographs spraying on to a continuous belt, which carries the twin tiles through an engobe spraying unit, then, where applicable, to an under-glaze spraying section and finally to the glaze section. The tiles will then automatically turn over so that the side which was facing away from the airograph during the first operation becomes the uppermost surface, and is treated to exactly the same sequence as the first tile. From the glazing area tiles are dried again, placed on their edges on kiln cars and once fired to produce a vitrified and glazed tile. According to the body and glazes used it may be necessary to coat the edges of the tiles, which support those above them in the stack, with a bat wash mixture to reduce the chance of their sticking during firing.

When the tiles are fired they are sorted according to size – large, average and small – and packed into boxes each usually containing a square metre. This variation in size is due to irregular water content in the clay mixture in the pugmill, and occasionally some inconsistency in the blending of the body or the temperature in that area of the kiln where the tile was fired. When set, the variations in size are not noticeable. The tiles may be split before packing or on the site where the tiles are to be set. There is some advantage in leaving the splitting until the tiles have reached the site as, back to back, the two tiles are stronger and less subject to damage in transit than single split tiles. Splitting is carried out in a simple wooden frame which holds the tiles on edge, where they are separated with one blow of a hammer on a wedge or bolster.

These tiles are most successful when set under factory conditions into pre-cast concrete units. The system is to lay the tiles within the frame for casting the concrete unit, then to pour the concrete, locate the reinforcement and back up. Such panels may be transported to the building site and erected quite simply on steel frame structures, providing a durable, weather-resistant surface which is richer in its material qualities than the raw concrete.

Other shapes may be extruded, some of which bear close relationship to hollow brick extrusion, particularly

82 Splitting extruded clay floor tiles with a bolster and hammer. One light, sharp blow should be sufficient.

when the tile takes the form of a moulding with perhaps only one surface glazed. In a studio it may be possible to undertake the extrusion of small quantities of special tiles and mouldings, but the problems of handling, after the clay passes through the extruding die, require sophisticated engineering. While, in principle, this method of production seems simple, every large-scale manufacturer of extruded tiles is very reluctant to demonstrate how the problems of handling have been solved, as this may be the key to commercial success.

Shaws of Darwen, Lancashire, have recently developed a process for extruding a very thin tile. Conventionally twin tiles are between 9 and 13 mm thick including the dovetail, and this physical characteristic means that the tile cannot be set without the services of a skilled tile-setter. Such people are increasingly difficult to come by, as the market, in the UK at any rate, is characterized by large sales in the do-it-yourself area. To meet the demand of this market, tiles must be thin, easy to carry, and lend themselves to fixing with modern adhesive techniques, and not the traditional sand and cement setting. The thin extruded supertile may not replace the heavier, conventional twin tile as a flooring material because the demand for floor tiles is largely determined by building in the public sector. For domestic flooring, tiles are not as popular in Britain and the USA as they are in Italy, France and Spain.

CAST TILES

Slip-cast tiles are commercially produced by casting large tiles in banks of moulds where the pouring holes are interconnected, as are the risers which allow the air to escape. Otherwise slip casting would only be used to produce those surfaces which would be unobtainable in any other way.

The tiles should be produced as a clay or plaster model and from that model a mould should be made. The mould is made of plaster of Paris, usually in two pieces which separate along the back edge of the tile. The top of the mould, which is the back of the tile, contains one or more pouring holes, according to the type of relief on the surface of the tile. To produce the mould the model is placed face uppermost on a glass surface and cottles are placed 50–75 mm away from the edge of the model. The plaster is mixed in the correct proportions with water, blended to ensure that there are no lumps and poured on to the model, casting around and over it to produce a thickness of plaster above the model about 50 mm thick. If the model itself is made of plaster it must be sized, or impregnated with resin, to prevent the plaster from sticking. The modelling on the surface must be such that it does not form any physical keys with the plaster mould, otherwise the model would be impossible to remove.

83 Faience tiles cast in banks. The moulds are banked as much as twenty high and the tiles are cast by pouring slip down one hole and allowing it to rise through the opposite side, filling each mould as it does so. As the moulds are dismantled each tile is scored with the wide-toothed comb.

When the top and side surfaces have been cast the model and mould are turned over, the sides of the mould, which have just been cast, soaped, and a clay or resin-impregnated plaster plug placed in the centre of the back of the tile. Fresh plaster is mixed and poured on the back of the model to the same dimensions as the first part of the mould. When the plaster is set the mould is separated and the model removed. It is important that the first part of the mould has notches cut into it to allow the second piece, as it is poured, to run into these, producing registration devices which will allow the mould to be assembled correctly.

When the plaster has dried it is filled with a deflocculated slip and cast up to the appropriate thickness to produce either a hollow cast or a solid cast, according to the thickness of the tile. When the slip is cast to the correct thickness the mould is drained and left to dry until the cast tile is in a state where it may be removed without damage. Any seams which are evident on the cast piece should be fettled off and the piece left to dry. After which, it may be bisquit-fired, decorated and glazed in the normal way.

This process is probably most suitable for tiling units with complex surface characteristics, which cannot be pressed or extruded to give adequate definition. It may also be suitable for large single tiles where the scale is such that a hand-pressed piece of clay would be almost impossible to dry and fire without warpage. In the latter case it may be necessary to add grog to the slip to aid drying and to inhibit the collapsing of the tile during the firing process.

84 Fly press, operated by spinning
the flywheel (top), which lowers the
top die by means of the thread on
the central column. The dust clay in
the bottom mould is compacted by
the top die.

PRESSED TILES

Anyone who has fired clay tiles, made from plastic clay,
will be aware of the difficulty of achieving a flat finished
tile. During the firing and drying the tendency is for the
moisture to leave the clay unevenly, and although this
process may be carefully supervised and care exercised in
the drying and stacking, it is not possible to guarantee the
final quality of each tile. By 1840 attempts had been made
to produce clay objects with a minimum amount of water
by compressing clay dust between two metal dies. This
dust contained between 10 and 20 per cent of moisture as
opposed to the 30 or 40 per cent necessary in the manufacture of plastic pressed tiles. Clay objects demonstrate a
'memory' of the process by which they have been formed,
and nowhere is this more obvious than in the production
of tiles. The procedure of rolling clay out tends to produce
a curved slab rather than one which is flat. This can be
attributed to the varying pressures concentrated upon the
tile at different times. The pressing of dust clay does more
than reduce the amount of water required to make the clay
flow to the desired shape. The pressing is carried out very
quickly and if the dies are true, with the correct amount of
powdered clay placed between them, the tiles produced
will be of a standard thickness and density. Thus the major
problems associated with plastic pressing are eradicated at a
stroke. Needless to say, this easy production cannot be
achieved without some experiment to determine the most
satisfactory body for dust pressing, the minimum amount
of moisture necessary to create a bond between the particles
of dust, and the ideal amount of dust to produce a tile of the
required thickness.

There are several types of press for the production of tiles.
The simplest is the hand-operated fly press, which consists
of a heavy cast frame which supports the bottom die about
one metre above floor level. The frame is surmounted by
twin columns which hold a large threaded screw carrying
a cast flywheel, and to which it is fixed. At the bottom of
the screw is fastened the top die. The top and bottom dies
are secured in place so that as the top die is lowered it will
slide easily into the bottom die without touching the sides
and damaging either part of the die. This setting up must
be done with considerable care, as the slightest inaccuracy
may cause severe damage to a very expensive machine.
The bottom die consists of two parts: a movable bottom
plate and a frame which is fastened at each corner to the
bed of the tile press. The various parts of the die are fitted
to the press in the following order. The bottom plate is
screwed on to the movable ejection lever. The tile frame is
clamped on to the tile press bed and registered with the
bottom plate. This is relatively easy as the frame will only
fit one way once the bottom plate is in position. However,
the bottom plate should move easily up and down, so

some slight adjustment to the location of the frame may be necessary to ensure this ease of movement. The top die is then screwed on to the top thread but not fully tightened. It is then lowered by slowly moving the flywheel. The flywheel tends to accelerate as it gathers momentum, so it must remain under some restraint until you can ascertain the position of the top die in relation to the bottom frame.

It is more than likely that the top die may be out of alignment as much as 45° and, in order to correct this, the top die must be backed off by that amount. This is only possible by lowering it on the threaded top screw, which means that the top die would be loose and free to rotate as it is lowered during the pressing operation. Therefore, the space between the stop on the die and the top screw must be packed with metal spacers or washers, so that when the die is in the proper relationship to the frame it is also tight on the top screw. Time, and a little patience, may be required to ensure a satisfactory alignment. When removed from the press the top and bottom die may appear to be identical, apart from the grooves machined into the bottom die. Closer inspection will reveal different sizes of thread where they are to be fitted either on to the lower ejection lever or on to the top screw to which the flywheel is attached. This difference in size of thread is quite deliberate, and ensures that the dies are installed in the proper manner.

The grooves on one of the dies are intended to produce a key on the back of the tile so that a sound fix will be produced when the tile is installed in its final setting. If on the lower die they will be so designed that the tiles, when pressed, will slide across the die and on to a carrying board, so that they can be removed from the press without damage. Once the die has been assembled and aligned the press should be operated several times slowly to ensure a smooth action, and a few pressings made to determine the amount of clay required for a specified thickness of tile. The degree of pressure and number of pressings required will be important factors in determining the final thickness and density of the tile.

It is not unusual in hand pressing to press each tile twice; once by letting the die come down under its own momentum, and again with some acceleration provided by the operator. The first press squeezes the air out of the powder and should be carried out quite slowly. If the dies have been correctly designed there should be enough space between the tile frame and the top die to allow the air, included in the loose clay, to be evacuated in the first pressing. The second pressing compresses the powder to the desired thickness and density. This pressing operation is quite hard work if many tiles are to be produced. The top die must be raised at the end of each pressing in order that the finished tile may be removed. This means that the flywheel must be raised by hand against the natural gravitational feed on the thread.

The top die will include either a ball bearing or a steel hemisphere which acts as a stop button for the bottom of the top screw. As the die is screwed tight to the top thread the button is clamped between the thread and the die. This button has the effect of causing the die to bounce when it is lowered on to the clay so that the operator has only to spin the flywheel and catch it as it starts to return. The whole operation takes only seconds and one operator working with an assistant to remove the tiles can produce more than 3,000 tiles per day.

Tile presses are very dangerous pieces of equipment and must be used with considerable care. Hand-operated fly presses may not have the kind of guard associated with automatic presses and it is possible to trap your fingers in the die, either by leaving your hand in the press while the top die is released, or by not securing the locking device normally provided to hold the flywheel in position when it is fully raised. Under no circumstances should two people operate the press simultaneously as, sooner or later, one will start to lower the die while the other is still filling or levelling off the dust. Some presses have handles projecting down from the flywheel and care must be exercised during the pressing operation; for as the flywheel turns and moves under its own momentum the handle may strike anyone standing near by, or the operator himself.

Dies for pressing

Dies supplied by the manufacturer of the tile press are normally made of hardened steel, which is very expensive

85 Two etched dies (left) and the tiles produced from them.

86 Automatic tile press with the dies disassembled. The centre die shows that the top face can be replaced when worn.

87 Automatic tile press designed to press four tiles simultaneously. In use, this press would feed a conveyer system which carries the tiles away to be fettled and stacked ready for firing.

but long-lasting. For studio production the dies may be made of various materials provided that longevity is not essential. Often only the top die need be made, if suitable bottom die shapes are available. If unique shapes are required, it would bring into question the idea of using a dust press, as dies take some time to perfect. Wooden dies may be the easiest to make: these may be fastened to the press by fixing a bolt with a thread matching that of the top and bottom threads of the press. If a new frame is to be made this should be shaped to accommodate the fastenings already on the press bed. Glass-reinforced plastic dies may be made by casting, and these have the advantage of allow- ing the casting of the necessary bolts into the top dies, which will give an excellent fix.

Simpler modifications may be made to the pressing services of the top or bottom die. Plates may be acid-etched, cast or engraved, and placed inside the frame so that they rest on the bottom die and require no permanent fastening. Any modifications to the top die, however, must be firmly attached; the simplest method is to use a water-soluble adhesive, so that the plate may be removed by soaking the top die in water. (After this, it is essential that the die be thoroughly dried and greased to prevent corrosion.) Dies which are to be used for production runs, or even

88 A cast metal die for producing dust-pressed embossed tiles.

several hundred tiles, are best made up from metal, but it may be easier to fabricate these rather than to machine solid blocks of steel, which will require sophisticated machine tools.

Industrial dust presses conform to the basic design of the hand presses, but are of two types, semi-automatic and automatic. A semi-automatic press looks similar to a hand press, but the operator only fills the die with dust, the pressing operation being done automatically when the hand guard is lowered into place. The automatic press is loaded with the correct amount of clay, levelled off, pressed, raised, and moved on to the conveyer without any manual operations being involved. The dies are heated and sprayed with light oil or paraffin between each pressing, so that the tiles may be released with ease and there is no chance of the clay sticking to the dies.

Moisture content in dust for pressing

While the moisture content in clay dust may vary owing to atmospheric conditions it should be brought to between 10 and 20 per cent before use. This is less critical in the case of hand pressing, where the process is less regulated, than in semi- and fully automatic pressing. Considerable efforts are maintained in tile factories to ensure a constant and regulated moisture content in the dust. The body is prepared in large arks containing the various ingredients which will give it the required characteristics in the making as well as the fired state. This slip is pumped to the filter press and the cakes produced are then dried and crushed. More recently, spray driers have been used, where the slip is sprayed into a tall cylinder under pressure and instantaneously dried as it meets a stream of hot air simultaneously pumped into the chamber. This has the advantage of keeping the particles of clay separate and avoiding the necessity of caking the clay only to crush it up again. From the crusher, or drier, the clay is taken to the press hoppers by conveyer, and delivered to the press in regulated quantities. During the passage to the presses the clay is periodically tested to ensure that the water content is kept within acceptable margins. Any variations beyond those limits mean that the dust-making process is interrupted until the water content has been adjusted – which will cause a considerable loss of production.

Automatic tile presses are sometimes designed to press several tiles at once. Hand and semi-automatic presses may also use dies to produce several tiles at a time, but these will probably be small, perhaps no more than 2 in. (5 cm) square. Dies may be designed to produce tiles with spacer lugs to simplify the setting of the tile and obviate the highly skilled setting with spacers when the tiles are installed. The markings on the reverse of a clay tile are intended to provide an adequate key for adhesive fixing, which has been developed in the last twenty-five years. Where these key grooves run in only one direction it is an indication that the tile has been pressed with these markings on the bottom die. If the keys cross over each other to produce some kind of lattice it is likely that they were produced by the top die, for otherwise it would be impossible to slide the tile from the bottom die.

Given that the clay is in suitable condition, pressing tiles is relatively easy and does not require particular skill. With the dies fixed correctly, the bottom die is pressed as far as it will go and the cavity filled with clay. To ensure even filling a levelling stick is run over the dust, and this action both evenly distributes the clay in the die and removes any excess. When the cavity is filled and levelled, the flywheel is released and the top die brought down on to the dust, compressing it and squeezing out any air. The top die is raised by turning the flywheel in the opposite direction from that in which it turns of its own accord; it is then locked in position, if a lock is provided, or held in position with a suitable piece of chain, hooked on to a rigid part of the framework and to a convenient point on the flywheel. The tile is removed by pressing the foot lever or hand lever, if there is one. This raises the bottom die and brings the pressed tile above the level of the tile frame. It is a relatively simple matter to slide the tile in the direction of the bottom grooves, and on to a suitable carrying board. The tile should be checked for thickness and regularity, i.e. that there are no damaged corners or edges. If the tile is too thick, less clay should be put in the press; to ensure a standard thickness each charge of dust should be weighed once a satisfactory weight of clay for a given thickness has been determined. An easier method of regulating the quantity of dust is to use a standard scoop. If the tile has crumbled at the edge, or corners are not tightly pressed, the clay is probably short of moisture. Either the tile should be pressed twice or the speed of the flywheel should be accelerated to increase the force with which it presses the dust. This may be a satis-factory solution if only a few tiles are to be produced, but if a long run is intended, it is worth while to make the extra effort to modify the water content of the clay as this will be physically less demanding in the long run.

After several pressings you may find that the die starts to pick up dust from the tile as it is pressed. This is likely to occur at the most susceptible parts of the tile, i.e. the edges and corners, and the centre of the top face. It is common practice to wipe the die with an oily rag after each pressing and this lubricant between the clay and the die will prevent sticking without affecting the tile surface. Alternatively, paraffin may be sprayed on to the top and bottom dies.

After pressing, the tile should be set aside to dry before placing it in the kiln. Usually a few hours in a warm place will suffice; the most convenient place in the studio is above

89 Levelling off a dust-filled die.

90 Raising the tile after pressing.

or beside the kiln. The tile may then be carefully removed from the carrying board and stacked in the kiln.

Pressing with a semi-automatic press follows much the same procedure but the water content of the dust is more carefully controlled, and the operator has a measure which is filled with the correct quantity of dust and struck off with a steel level. This whole procedure takes only seconds, and is much faster than weighing each charge. The dust is tipped into the die and a guard brought down to protect the operator from the action of the die and flywheel. This guard also operates a hydraulic system which releases the flywheel and brings down the top die at a predetermined pressure. The die is then raised hydraulically and the guard lifted. The bottom die may then be raised and the pressing removed.

It may be necessary to fettle the edges of the tile after pressing, and this may be carried out with a slightly damp sponge or soft leather. At this point the tiles are quite fragile and you should be very careful to keep any losses, due to careless handling, to an absolute minimum.

Many of the procedures described earlier in this chapter as modifications to a plastic clay-making technique may also be applicable to dust pressing. Small templates, cut from card or rubber, may be fixed to blanks inserted into the bottom die. When the dust tile is pressed the tile and blank, together with the template, may be removed and turned upside down, and the blank and template taken away to reveal a relief tile. Many simple techniques may be adopted to produce one-off tiles, and in cases where the design looks particularly promising more permanent dies may be made for prototype or production runs. The clay body used to produce the tile need not be of a single colour. Various marbling and speckling techniques may be generated by placing powder, or plastic clay, of contrasting colour into the die before packing up the conventional tile body.

Plastic pressing

Plastic clay, which is slightly softer than leather-hard, may be pressed in fly presses with greater ease than is apparent with dust clay. The clay should be knocked up into a rectangular block and sliced into layers with a cutting frame. The precise thickness of each layer must be determined by experiment so that when the layer is set into the tile die and pressed, it will produce a tile of adequate thickness and density. Plastic pressing of this kind lends itself to all sorts of inlay and relief pressing. If the clay is lightly pressed at first, in order to shape it into the die, a secondary die bearing a relief image may be placed on top of the clay and pressed home using slow but firm pressure on the top die. In this case it is advisable to use a release oil, such as rape seed oil or diesel oil, to ensure that the metal plate used as a secondary die may be easily removed from the clay tile.

Faults in pressed tiles

The most likely faults to occur in the process of pressing tiles are clay sticking to the dies, the tile crumbling when removed from the die, and irregular thicknesses revealed when several tiles are put together. Cures for sticking have already been mentioned, as have the likely reasons for weak tiles. Irregular thickness can be overcome by careful measuring of the amount of clay loaded into the die either by weighing, or fabricating a container which will hold the measured volume of clay when it has been filled and struck level with a wood or steel straight-edge.

One fault which occurs from time to time is revealed by the central area of the top surface of the tile becoming separated from the rest of the tile. It might be assumed that the point of maximum pressure upon the dust would be directly beneath the threaded column carrying the top die. In fact this is not the case: the forces acting upon the clay powder pass through the material and bounce back from the walls of the frame, resulting in a tile with a more lightly pressed central core than the outer edges. This can be overcome by modification of the various particle sizes making up the body, or more easily by adjusting the water content in an attempt to improve the bond between the particles of clay.

If the tiles are found to have very weak corners it is likely that this is due to careless loading of the die when the dust was being introduced, or to small amounts of clay being used in an attempt to produce a very thin tile. With hand pressing, standardization of the water content in the body and pressing the minimum number of times to achieve a satisfactory tile are the most likely ways of avoiding the worst of these problems.

11 Decorative tiles

There are many different ways of producing decorative tiles and this chapter is concerned with those methods of decoration which are closely integrated with the method of forming the tile. For thousands of years it was possible to make tiles only with plastic clay, and it is on the qualities of the clay in this state that many of the decorative processes were developed.

IMPRESSED PATTERNS

The surface of the tile may be modified by hand modelling to produce a decorative surface. The simplest technique involves repeatedly pressing hard objects, such as modelling tools, into the clay tile, so that a systematic pattern is produced. Any object may be used and it is not necessary to define the process more specifically as the possibilities are infinite. Large or small, natural or man-made objects may be pressed into the tile once only, or repeatedly, with a clearly defined impression or overlapping to give a profusion of indentations in the clay.

Whenever you press into the tile surface it is advisable to keep the tile either in the mould in which it was made or in a frame made specially for the purpose if insufficient moulds are available. The tile may be removed as soon as the pressing operation is complete, but should it be pressed without any support for the sides it will almost certainly be pressed out of shape. This distortion may be sufficient to destroy the over-all shape of the tile, or it may be less apparent but enough to disrupt the setting of the tile on site. Even when a mould or frame holder is used it may be necessary to fettle or trim the edge afterwards if the impressions are close to the boundary of the tile.

If the object tends to stick to the clay during pressing this may be because its shape is such that it locks into the clay, or else it presents such a large surface area to the clay that a vacuum is created as you attempt to remove it. Such a vacuum may require such energy to overcome it that the tile may be destroyed or distorted beyond repair. If the object is locking into the tile because of its shape either the object should be modified so that it will release, or it should be discarded. If the locking is caused by vacuum this may be due to the surface being water-absorbent. Making it

91 A cast metal die for a relief tile, with tube line effects.

non-porous may not be enough so a coating of oil or petroleum jelly should be applied which will at least allow the object to slide out of clay. It may also be that the clay is too damp for easy pressing and should be allowed to dry before continuing.

STAMPED TILES

Rather than use modelling tools or found objects to create the design on a plastic clay tile, you may wish to create a more regulated image. This can be done by making a stamp, of wood, plaster or bisquited clay. Whichever material is chosen it should be cut to the size of the tile and the required image then cut into it. This process is most suitable for plaster stamps, or clay which should be fired before use, making allowance for the shrinkage during firing. If the stamp is to be of wood the pattern should be cut from a piece of a suitable thickness and fastened by nails, screws or glue on to a wooden blank, incorporating a handle if necessary.

Once the stamp has been produced it may be used exclusively on the tiles to be decorated, but an interesting variation can be achieved by casting against it a plaster of Paris female stamp. These two stamps, one negative and one positive, may be used to produce identical images but different surfaces. Wooden dies may also be made by carving into the face of the stamp to produce the desired image. Stamps such as these have a long tradition in the history of tile decoration and examination of any collection of such tiles demonstrates the ingenuity which tile-makers have brought to this technique.

The stamping technique for decorating tiles includes the use of rollers. These can be made quite simply by casting cylinders of plaster of Paris and treating the surface of the roller by cutting and modelling to produce the desired image. When rolled across the surface of a plastic tile the clay will accept the impression made by the roller. Such rollers, with small diameters, may be rolled across the tile to produce repeating patterns, and if the engraving on the roller is carried out with sufficient care there may be no breaks in the repeating pattern impressed upon the tile. Rollers are still in use commercially, particularly for decorating pottery after it has been turned and while still in a leather-hard state. Such rollers may be made of metal and can be bought for use in studios.

Any plaster of Paris roller or stamp may be strengthened, and its working life thereby increased, by impregnating the plaster with epoxy resin to produce a hard, non-porous and very strong surface, which does not break under the repeated pressure needed to produce the image on the clay tile. Such resins are available commercially and are supplied with data sheets explaining precisely how they should be used. During the impregnation process toxic fumes may be given off, therefore this should be carried out in a well-ventilated area and a respirator should be worn. Always inform the supplier of the resin of the purpose for which you wish to use it and follow his advice and instructions. The resin-impregnated dies will need to be heated to cure the resin and ensure that it has impregnated the plaster instead of remaining on the surface. It may be necessary to use a thermostatically controlled curing oven.

If it is not possible, or desirable, to impregnate a plaster stamp with resin you should consider using it as the bottom surface of a tile frame into which the clay is pressed. Such frames should be much deeper than those used to mould a flat plastic tile: if the plaster die is 2 in. (5 cm) thick the frame should be $2\frac{1}{2}$ in. (6.3 cm) higher to produce a tile of $\frac{1}{2}$ in. (13 mm) thick. Plaster technology has improved considerably in recent years, and you may find that plasters which set very hard will not require resin impregnation to give sufficient strength to the plaster for your purposes. These have the advantage over resin-impregnated dies, or stamps, that they retain some degree of porosity which assists in releasing the clay from the plaster surface. During any stamping or pressing operation you should inspect the plaster die to ensure that it is neither wearing too rapidly nor breaking up. Impregnated dies will not break up or wear as rapidly as untreated plaster, but as clay is acidic and abrasive some wear will occur, and if this passes unnoticed you may produce tiles with a faded image which will not match tiles made when the die was younger. If long runs are to be made you should consider whether or not a master pattern should be kept, so that if your die wears too quickly it may be replaced by a new one cast from the same master, thus ensuring the pressing of identical patterns.

It is not possible to produce satisfactory dies or patterns involving undercuts, nor is it advisable to create very sharp changes in plane. Undercuts will prevent stamp and clay from releasing, after the pressing has been carried out. Sharp changes in plane will wear rapidly as pressure will tend to concentrate on these areas and the clay will have to flow and shear to produce the identical pattern. All this will rub away the crisp modelling and the stamp may have such a short working life as to be unsuitable for its purpose.

ENCAUSTIC TILES

Some of the earliest known ceramic tiles were decorated by inlaying coloured clay. It is likely that the pattern was cut into the plastic clay and that clays of a contrasting colour were pre-shaped and pressed into the cavities in the tile. Truly encaustic tiles seem to have been an invention of western Europe and by the thirteenth century had achieved a richness of material and imagery which remains unsurpassed. These tiles were used particularly for flooring as the image deteriorates only after considerable wear.

Encaustic tiles are traditionally made of red-burning clay pressed into a tile frame and stamped with an image which produces recesses in the surface of the tile. This series of depressions is then filled with white clay (traditionally pipe clay) in a liquid or very soft state, so as to overfill the hollows. The tile is left in the frame to harden to a leather-hard state, when the white clay is scraped or planed away to produce a flat tile with a white image integrated into the body.

Obviously the impressed image may be built up of several die stamps where this is more convenient. It is possible to extend the apparent printing technique so that the image is produced by several impressions, each in its turn filled with a clay of a different colour. The major problem in the production of encaustic tiles from plastic clay and slip is that as the tile body and the image contain different proportions of water, the inlay has a tendency to shrink away from the tile body. If this shrinkage is excessive the inlaid image may drop out after firing. Even if the inlay and the tile have the same thermal expansion this problem may not be overcome.

Several cures are possible. Suitably coloured grog may be added to the slip and will have the effect of reducing the shrinkage. Such inclusions of grog should be finely ground so that during the planing operation no unsightly scars are caused by large pieces of grog being dragged across the surface of the tile. Alternatively a deflocculant may be added to the slip, which will allow it to flow with a reduced water content and therefore a reduced shrinkage through the drying and firing cycles.

The deeper the impression and inlay the greater will be the problems of shrinkage. In order to avoid the worst excesses of this fault the tile-making clay should be relatively plastic, so that it shrinks on to the inlay, and the tile should not be too dry when the slip is poured into it as this will inhibit the creation of physical bonds between the two different clays. After the image has been made and cleaned, the tile should be removed from the frame and trimmed to an accurate shape. During this process any slip which has run on to the edges of the tiles may be removed.

An alternative method is to produce an open-topped mould, with the image intended for the face of the tile in relief on the bottom. The clay is cast as a thin layer, allowed to stiffen, and then backed up by pressing plastic clay into the mould. When leather-hard, the tile is removed and the hollows on the face of the die filled with clay of a contrasting colour. When this too is leather-hard the surface of the tile is planed level. Some nineteenth-century tiles made by this type of process are laminated with the clay on the top of the tile repeated on the bottom. This would have the effect of reducing the warping of the tile due to differential shrinkage of the surface clay and the tile body clay.

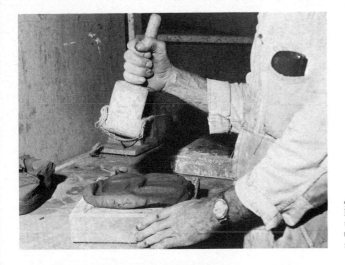

92 Pressing encaustic tiles. The clay has been wedged and is being tamped into the mould. The tamping tool is covered with hessian so that the clay does not stick to it.

93 The final pressing is continued by hand, overfilled and tamped down again.

94 The tile being cut level with a cutting wire.

95 Levelling the back of the tile with a steel blade.

96 Spiking the tile to provide a key for the mortar and to facilitate drying.

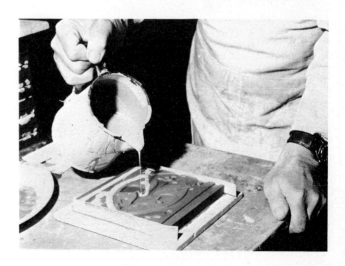

97 The tile removed from the mould. Any open recesses are sealed off with damp paper; this sticks to the clay and the engobe inlay is poured into the cavities.

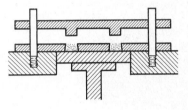

98 After drying, the excess engobe is cut away with a plane, blade or similar tool.

Dust-pressed encaustic tiles were made by producing the inlay as a separate pressing, locating it in the bottom of the press and then backing up with tile dust. Several methods of shaping the coloured inlay were used, from bent wire frames soldered together in the shape of the image to pierced metal plates. In every case the inlay was first lightly pressed, the frame removed, the inlay put in the die and pressed. While this technique avoids the problem of the inlay shrinking and the tile warping it is a slow process, and there is a risk that the inlay will be damaged or spread under the pressure of the tile-pressing operation. This might be overcome by the use of resins and waxes to harden the inlay but plastic and dust-pressed tiles continued in production into the twentieth century without either technique gaining a monopoly.

While the common material for making the die for encaustic patterns was initially wood, various metals, plaster of Paris and other materials have been used, particularly during the nineteenth century, when encaustic tiles were produced in great numbers. The harder the material the longer it would last the constant pressing, particularly with the use of mechanical presses. In the studio, however, where only short runs are envisaged, almost anything, from card to pieces of linoleum, may be used. The latter can be cut with relative ease, leaving the fabric backing to maintain any separate parts of the image in the position required to produce the necessary design effect. Many trials may be made and patterns produced using the simplest means very quickly. The only problem which may occur is the difficulty sometimes experienced in removing the stamp from the clay tile, but grease or oil will solve this problem.

If the tile is to be decorated as a flat tile it would be sensible to leave the decoration process until the tile has been fired at least once. This will reduce the risk involved in spending time on a piece which might contain a fault revealed only

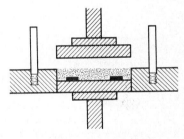

99 Dust-pressing encaustic tiles. *Top:* The bottom die of the press is raised and the pattern frame located on the pins set into the bed of the press. Coloured granules of clay are set into the pattern plate and pressed on to the bottom die, using the relief plate which is also located on the pins in the press bed. *Bottom:* The pattern and relief plate have been removed, and the bottom die lowered and filled with powdered clay to contrast with the pattern. The whole is pressed by the top die to compact the clay dust and form the tile.

100 'Indestructible lettering' in black and white glazed encaustic tiles, from the Minton catalogue of 1918.

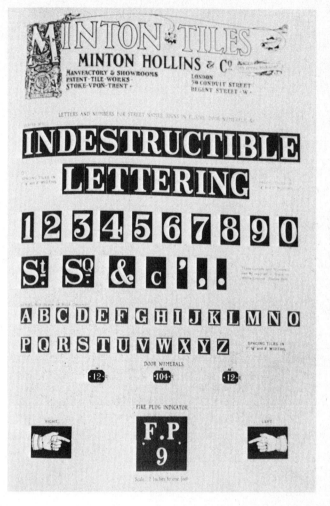

101 Section of an encaustic tile. The top surface is matched by a similar layer at the bottom, which reduces the tendency to warp during drying and firing. The thicker central layer is of coarser clay, unsuitable for this kind of decoration but providing the necessary physical strength and drying qualities.

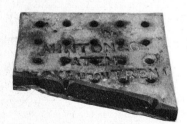

102 The reverse of an encaustic tile, spiked to improve drying and fixing.

when it is fired. Some processes lend themselves to both fired and unfired tiles, the only difference being the composition of the material being applied.

SLIP TRAILING

Slip trailing entails the application of an engobe to the tile by means of a slip trailer. In this context the engobe may be a different colour from that of the tile so that a raised contrasting line is produced. Where the tile body is very light in colour it would be best to use this as the basis of the engobe. Unfortunately, most white-burning clays are too plastic, and therefore shrink too much, to lend themselves to tile production, although modification by calcining some of the clay, or the addition of grog, may reduce the shrinkage and warping which goes hand in hand with high plasticity. The result might be that the tile and engobe have differing rates of expansion and contraction during drying and firing and they would probably separate. The engobe may fall off the tile, totally or in part, leaving the pattern incomplete.

This type of tile is made most commonly of red-burning clay which can be compatible with white clays. The engobe is made by adding water to the clay until it is of a creamy consistency. This may be done using a powdered or plastic clay. While it is easier to achieve a smooth suspension with powdered clay it may be found that such engobes have slightly different physical properties as they age. They may become more plastic and run more easily. Engobes made from plastic clay may have these advantages immediately they have been mixed, or at least develop them very quickly. For the use to which engobes are put these differences are slight and may pass unnoticed.

TUBE LINING

This technique is similar to slip trailing in that the edge of the design is produced by trailing a very fine line of engobe on to the tile while it is still green. The trailed line may be the same clay as that of the tile, as a contrast is not always required, but the line should be as sharp as possible. The clay used to make the engobe should be of a very fine particle size, and the outlet for the slip trailer should be about $\frac{1}{16}$ in. (1.5 mm) in diameter. When the engobe, with the necessary qualities and colour characteristics, has been produced, the trailer is filled by evacuating the air by squeezing the bulb, and immersing the tip in the engobe. As the bulb springs back to its original shape the vacuum created within it will then make the engobe enter the bulb. It may take a little while to fill, and 0.5 per cent of deflocculant may be added to the engobe to produce more liquid suspension, without increasing the water content. The deflocculant should be a solution of soda ash and sodium silicate, dissolved in a small amount of hot water. The total weight of soda ash and sodium silicate should not exceed 0.75 per cent of the dry weight of clay in the engobe. If this is exceeded there is a danger that the deflocculant will produce an effect the reverse of what is wanted. Should this 'flocking' effect occur, more clay must be added to the engobe to restore a balanced suspension.

When the bulb is full the fine nozzle can be fitted. Slight pressure on the bulb will produce a fine trail of engobe which remains as a raised line and does not flow over the surface of the tile. If the engobe blocks the nozzle it may be necessary to sieve the suspension so that any lumps are broken up or retained in the sieve.

The design may be pounced on to the tile or painted with a solution of vegetable dye and water. These outlines

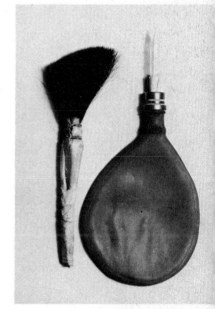

103 Tools for tube lining. *Left:* Glazing brush. *Right:* Tube liner, similar to a slip trailer but made of very soft rubber with a drawn glass nozzle to produce a fine line of engobe.

are followed with the trailer to produce a raised line, which will dry quite quickly as the tile absorbs the water in the engobe. Any cracking of the trailed line during the drying stage may not put a stop to the process of making the image, but it is an indication that the engobe is of too high a shrinkage. This shrinkage should be reduced in future compositions by adding less water, or including in the composition more refractory material such as calcined clay or grog, but these must be very finely ground. The design should be such that it includes enclosed shapes rather than free-running lines. This is because the purpose of tube lining is to contain the colours which are to be applied. When the lining is complete the tile should consist of a series of shapes surrounded by retaining walls of engobe.

The colour is applied either at the green or the bisquit state. In both cases it is applied under the glaze, having been made from metallic oxides or commercial under-glaze colour. The colour should be prepared by suspending it in a solution of gum arabic and water, which is then applied with a suitable brush to the appropriate area of the tile. Tiles covered in this way sometimes have the distinctive quality of variegated colour within the shapes. This is caused by the colour being held against the raised line during the application.

When all the colour has been applied the tile may be fired if it is in a green state. If the tile is already bisquited, or is to be fired once, the glaze should be applied by either spraying or dipping. The latter is possible only if the colour has been applied with a gum siccative. Without this ingredient the colour will tend to float off into the glaze batch, and thus contaminate any other tile similarly dipped.

This glaze should be transparent and craze-resistant. The final effect when taken from the kiln is that the tile has been decorated with glazes of different colours.

Tube lining is still carried out by at least one manufacturer of industrial tiles. The composition of the engobe is such that it may be applied to a bisquited tile, and it contains sufficient fluxes and refractory material for it not to shrink during drying and to fuse itself to the bisquit tile during the glaze firing. This use of vitrifying engobes allows for the decoration to be applied to the tile, which is then bisquited, filled with coloured glazes, and fired again.

An effect similar to tube lining was popular in the late nineteenth century when tiles were produced in mechanical presses using a die with grooves cut into it to produce lines which act physically in the same way as a tube-lined retaining wall. These tiles, however, are characterized by lines of regular thickness, unlike the hand-drawn, tube-lined imagery. The character of the tube-lined tile can be seen in the variation of thickness and height of the raised line. There may also be some flourishes at the end of some of the lines, depending on the type of gesture used when applying the engobe.

104 Tube lining: pouncing the pattern on to the tile with charcoal and a screen stencil.

105 The pattern pounced.

106 Trailing the tube lines over the charcoal pattern.

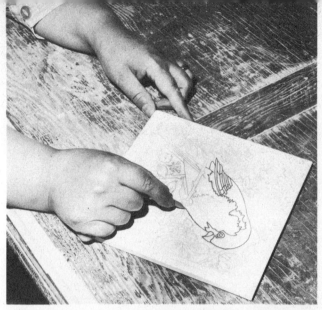

107 Filling the spaces between the lines with coloured glazes.

108 The glazed tile before firing.

109 The finished tile.

SGRAFFITO

Tiles may be covered with a contrasting engobe either before or after having been bisquit-fired. The engobe must be formulated so that it fits the body of the tile without crazing or peeling, so it must be a vitrifying engobe if applied to a bisquit tile; if the engobe is sprayed on to the tile, rather than the tile being dipped, it is possible to build up thin layers of engobe which can reduce the risk of a poor fit. Once the engobe has been applied, lines can be drawn through it with a sharp metal or wooden tool to reveal the body of the tile. Multi-toothed tools or combs may be used to produce several parallel cut lines, and by modifying the teeth of these combs lines of varying thickness may be produced simultaneously.

110 Monochrome tile with white engobe on red-firing clay, sgraffito technique; by M. L. Solon, 1877.

A reverse sgraffito may be made by scratching lines into a plastic tile, flooding these lines with engobe and then planing the surface, so that all the engobe is removed except that within the incisions in the clay. Such a tile is similar to an encaustic tile, but the pattern would be linear rather than containing blocks of contrasting colour.

Both these types of decoration may be repeated several times, using engobes of different colours to produce an image of greater subtlety and complexity. The greater the over-all depth of engobe the greater are the problems of fitting. Any variation in shrinkage or thermal expansion will accumulate with increasing thickness of engobe. If it has been made from the same clay as that of the tile it is likely that they will prove compatible.

ENGOBE DECORATION

Two types of decoration are possible using engobes in a liquid state. If an engobe is applied to the surface of a tile which is in an unfired, leather-hard state, another slightly thicker engobe may be applied, either in lines or spots, and these may be modified with a thin, flexible tool (traditionally the tip of a goose feather from which the hairs have been stripped without damaging the spine). Alternatively the engobe-covered tile may be spun on a whirler or tipped to and fro to produce whirls generated by the movement of the engobe across the tile. Great care must be taken not to add so much engobe that the tile absorbs water and becomes to soft to handle without damage.

WAX RESIST

Wax resist in the form of beeswax dissolved with paraffin may be painted upon the surface of a leather-hard or bisquited tile. When the tile is dipped into a suitable engobe or glaze, that area which has been painted with wax solution will not absorb water and, therefore, no engobe or glaze will adhere to it. The wax burns out on firing, leaving the body of the tile exposed in that area. This technique of wax resist may be put to practical use during any dipping operation to protect any part of the tile which is not to be coated by the suspension into which it is dipped.

While the wax resist technique works perfectly well when dipping into a suspension it is not so successful where contrasting engobe or colour is sprayed on to the wax. The action of the spray gun or airograph is to atomize the suspension and project fine particles of colour on to the object. There is no liquid flow across the surface of the wax and, therefore, particles of colour will adhere to it. If the spray technique has to be used on a wax resist area it is probable that the wax will have to be wiped with a damp sponge in order to produce a clean resist area. Alternatively, liquid rubber can be applied, and removed before firing.

PAINTING ON UNFIRED OR BISQUIT TILES

Painting a ceramic tile is relatively easy if you have any experience of painting on clay. A range of brushes designed for various purposes are on the market, and for painting on to unglazed tiles commercial colour or metallic compounds suspended in a solution of water and gum arabic give satisfactory results. Skill and experience are necessary to lay the colour on to the clay so that the brush strokes contribute to the over-all design and become an important feature. Only with considerable practice will it be possible to paint the design on to the tiles in such a way that the brush strokes are not apparent in the finished result.

If several tiles are to be produced with identical motifs, it may be necessary to pounce the outline of the pattern on to the tile before painting. This requires the production of a

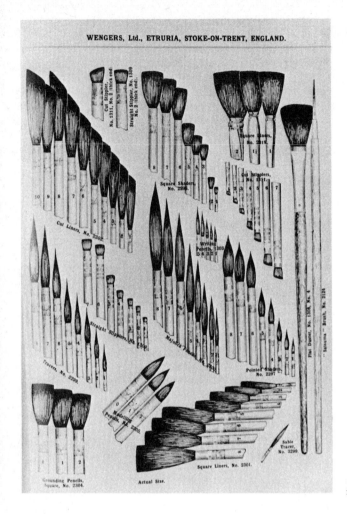

WENGERS, Ltd., ETRURIA, STOKE-ON-TRENT, ENGLAND.

111 Decorators' brushes illustrated
in a catalogue of about 1925.

drawn pattern upon thin paper, the pattern then being
pricked out with a needle, or a spiked wheel such as is used
in marking leather before punching holes in it. The transfer
is laid on the tile and charcoal or lampblack is dusted over the
paper pattern. When the paper is removed the dust which
has fallen through the holes produces a dotted line which
can be used as a guide when painting the outline of the
motif. Care must be taken not to move the pattern while
dusting on the powder. It may be easier to paint a weak
solution of vegetable dye over the pattern upon the tile so
that some dye passes through the holes. This latter method
reduces the working life of the pattern but is less susceptible
to error in registration.

Another type of guide is a transfer, or direct print on to
the tile, using a silk screen. Though charcoal and vegetable
dyes will burn out during firing, the transfer-printed outlines may be left as permanent images, since the painted

112 Two stencil-decorated tiles.

colour is normally of the on-glaze type. However, even when the colour and printed outline are both under the glaze the contrasting characteristics of hand-painted and printed colours are unpleasing to many tastes. This is partly because they were initially used as a cheap method of mass-producing outlines for tile-painters, and this quality of cheapness has become integrated into our aesthetic response.

Pouncing is not really a satisfactory method for producing a guide for on-glaze painting as both charcoal and dyes are easily dislodged from the glaze surface. It is useful for under-glaze painting when the pattern is transferred to an unfired or soft-fired tile, and when used on unfired glaze for in-glaze coloured effects.

Another method of repeating patterns on tiles is attributed to William De Morgan. The original was protected by a glass plate, on which the copy was painted. The glass was then removed and turned over so that the copy was on the underside and reversed as if in a mirror. Tracing or similar paper placed over the glass allowed the image to be copied in ceramic colours and when complete the paper was placed upon the tile and dusted with glaze. The paper burned away at an early stage of the glost firing to leave the pattern in place and under the glaze.

AIROGRAPHING

Colour may be applied with a spray gun or air brush. The colour is mixed with water, sieved through a 200-mesh sieve, and loaded into the airograph. The quality of line is determined by the closeness of the airograph to the tile surface: the closer the outlet, the thinner the line. The pressure of the air flow must be carefully controlled to ensure continual flow at constant pressure. Different types of airograph will require different pressures according to the particular job in hand. Commercially produced under-glaze colours are suitable for spraying, but the density of colour on the tile should not be such that it inhibits the development of a sound bond between glaze and tile surface, for then the glaze is likely to withdraw from the coloured areas, leaving them bare. This fault, when revealed in a finished tile, is known as 'crawling'.

As many commercial colours are pale they may be suitable only for application to white-bodied tiles.

Paper stencils can be used to mask areas of the tile which are to remain untouched by the colour being sprayed. Polychrome patterns may be devised which require only one firing with the glaze. Others may require a hardening-on fire before applying further colours.

DECORATIVE METHODS OF GLAZING TILES

Dipping

Traditionally, tiles were dipped into glaze so that only one surface was coated but parts of the edge inevitably picked up some of the glaze. The glaze is mixed in the normal way, sieved and brought to an appropriate pint weight. The surface of the glaze must be free from bubbles, and any which persist should be moved to the edge of the container to leave an undisturbed surface suitable for smooth contact with the tile. The tiles are picked up between fingers and thumb with the back of the tile uppermost. The glaze is applied by moving the tile through a short shallow arc so that it moves slowly across the glaze surface. Any marks left by your fingers may be repaired by fettling and applying drops of glaze to cover any bare patches. The finished surface should be even and free from variations in the depth of glaze. Tiles may be double- or treble-dipped in different glazes which react upon each other. This double dipping may involve re-covering part of the tile with one glaze and overlapping the second glaze to give a finished surface of three types, i.e. both glazes separately and the overlap area.

Airographing

Glaze may be sprayed on to either single tiles or several simultaneously. Images can be formed by spraying one glaze upon another which has been dipped on to the tile, but the image will have a soft edge and the double-glazed areas may react on each other. Stencils may be used to produce a crisper image, and if the first glaze has been applied by dipping it is less likely to be damaged by contact with a paper stencil than a surface which has been sprayed on the tile.

Techniques described under slip trailing, sgraffito and engobe decoration may be used with slight modification for devising decorative techniques for applying glaze.

Painting glaze

You may paint glaze upon the surface of the tile provided you take great care to ensure an even thickness of deposit. If you develop sufficient skill with this technique you will be able to paint many different colours on to a single tile without the necessity of firing each glaze before applying the next.

Droplet glazing

Machines have been designed, for use in tile factories, which spray glaze on to the tile with controlled atomization.

This means that several glazes may be applied consecutively, each glaze taking the form of large separate droplets, which when fired together produce a mottled 'all-over' effect. The origins of this technique lie in sprinkling glaze, or flicking it, from the end of a brush. Droplet glazing is open to much abuse as it requires very little skill, but needs a great deal of discernment if brash and insensitive effects are to be avoided.

While the techniques described in this chapter have been identified as independent methods of decorating tiles they may be combined in various ways. There are historical and contemporary examples of such combinations.

12   Printed decoration

It is not possible, within the scope and size of this book, fully to describe the skills and complexities of printing processes allied to ceramics. These are quite separate crafts, and few ceramists could hope to master all the skills of the printer as well as those skills which he needs himself. I shall limit the scope of this chapter to the ways in which these processes are applied to ceramics, and those features and possibilities which are peculiar to the production of printed imagery upon ceramics.

It is quite possible to devote a lifetime to ceramics and never use a printing process for surface decoration. Where a long production run is involved, or where the image itself, or the way it is to be used, requires a degree of standardization, printed imagery is essential. Certainly if any kind of photographic quality is sought a printing process is the only way it may be achieved. Designs may be produced as drawings or collaged. If the image contains any kind of printed material it may be necessary to reproduce this photographically in negative form, to make either a screen or lithographic plate capable of printing the image. If the image is drawn, it may be possible to transfer the drawing directly on to the screen or plate.

The differences between silk screen and lithography are that lithography uses a stone, or plate, which carries the coloured image in solid areas to be picked up by the paper when pressed against it; screen printing is a sophisticated form of stencil, in which those parts of the image which are not to be printed are blocked out, either with varnish or with paper, so that when the colour is pulled across the screen it deposits on to the surface beneath. The mesh of the screen depends upon the particle size of the colour. The finer the screen the better the definition, but in some cases, particularly in printing glazes, it is not possible to use fine screens, and the image must make allowance for this.

One of the major virtues of printed imagery in ceramics is that it enables polychromatic imagery to be put upon the ceramic with only one firing. Thus the value of the piece in commercial terms is considerably increased at comparatively little cost. It also makes standardization of decoration possible, which has been considered desirable by both manufacturers and customers.

Printed decoration may be divided into two basic types, with various subdivisions. These are direct printing, in

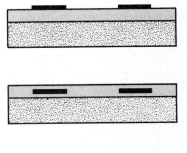

113 Location of printed images. *From top to bottom:* On-glaze, in-glaze, under-glaze (printed either on the bisquit or on the unfired clay).

which the image is printed directly upon the ceramic surface, and indirect or transfer printing, where the image is printed on to a suitable transfer paper, from which it is applied to the ceramic. Within these categories there are three types of possible location for the printed image on the ceramic: under the glaze, in the glaze and on the glaze. As these terms imply, the printing may take place at various stages of ceramic production.

DIRECT PRINTING

The most obvious application for this method of decoration is in the enhancement of tiles. The surface to be printed is flat, and may be located relatively easily in automated plant, where the tiles are produced and processed in linear system. Either glaze, or colour alone, may be printed. If the tile is to undergo several printings these may be carried out without firing each colour before applying the next, by using commercially available hardeners. These are normally fast-drying, so that the interval between printings may be no more than 10 feet (3 m) in length and, according to the speed of the production line, 30 seconds or one minute in time.

Colours or glazes to be printed in this way may be suspended in water- or oil-miscible media. If one print is made with an oil-based medium, and then glazed with a glaze suspended in water, the glaze will run off the printed image leaving it unglazed, so that the tile will have the appearance of the glaze having been cut away to reveal the colour beneath.

It is quite common to find even the simplest studio using a direct screen printing process. In the tile industry direct printing is part of the automatic production line, when all types of surface may be printed – under-glaze, in-glaze, on-glaze – and glazes themselves may be part of the decora-

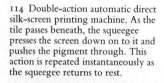

114 Double-action automatic direct silk-screen printing machine. As the tile passes beneath, the squeegee presses the screen down on to it and pushes the pigment through. This action is repeated instantaneously as the squeegee returns to rest.

115 Screen-printed tiles and
mouldings from Accrington station,
Lancashire (now demolished).

tion sometimes, finally passing through a waterfall glazing machine before being transported to the drier and the kiln.

Direct screen printing is the commonest method of printing for flat surfaces, although the Murray Curvex process is used in the tableware industry to print on to surfaces which have limited variations of surface, such as plates.

Many modern industrial tiles are 'cushion-edged', i.e. with curved rather than sharp edges. This reduces the area which can be direct-printed as the screen cannot print on to the curved edge. Any image designed for direct screening should normally allow for an unprintable 3 mm border on the edge of each tile.

TRANSFER PRINTING

There are many different types of transfer paper, each with its own qualities and each suitable for a particular process. Different papers may be used for lithographic printing and screen printing. The commonest, known as Simplex paper, is a water-absorbent backing paper with a thin printed gum on to which the image is printed. The image may be made up of several printings, each with a different colour. Each colour must be dry before the subsequent colours are printed, and both the printing and the drying must take place in a dust-free atmosphere, so a ceramic workshop is no place to produce printed transfers.

When the image has been printed and dried it is given an all-over printing of 'covercoat'. This should be printed to overlap the edges of the image. After drying, the image is ready for use. It is immersed in tepid water, which is absorbed by the backing paper. This expands and starts to curl as the water penetrates the paper, until it reaches the soluble gum beneath the printed image. At this point the paper flattens out because it is soaked right through, and there is no longer a variation in dimension between the back surface of the paper and that which is in contact with the print. The paper is removed from the water and the print and covercoat will slide from the backing paper – hence the term 'slide-off transfer' – taking with it sufficient gum from the backing paper to make it stick to the ceramic surface.

When the transfer is 'laid down' it must make perfect all-over contact with the ceramic in order to avoid unsightly faults when it has been fired. If the transfer is put on to a piece which has just come from the kiln, having been carefully handled to avoid fingerprints where the transfer is to be placed, no other cleaning should be necessary. If the object has been out of the kiln for some time it should be cleaned with a cotton-wool pad soaked in methylated spirits to remove any grease or dust. The spirit will evaporate quickly to leave a clean, dry surface ready for the transfer.

As the transfer is placed on the object it may be moved by sliding it into exactly the right position, and any air bubbles should be squeezed out with a rubber kidney, or fine cotton pad, working from the centre outwards. The transfer and covercoat are slightly elastic, which is useful if the form is curved, or has slight variations on the surface. If the transfer has been stored for some time it may become brittle and this will make it difficult, if not impossible, to apply without cracking or tearing.

Several transfers may be fired one upon the other, provided that each is fired before the next is put down. The only reason for putting on several transfers would be if you have not decided exactly what the image is to be before starting to put on the first transfer. It would not be a very profitable way of producing many identical objects, but it can be useful for developing ideas, particularly if you have several spare transfers from previous projects, or have some with a small printing defect that makes them useless for the original purpose. Occasionally, when working on a sculptural piece, a variation of surface image without too much change in surface quality may be desirable; using several parts of different transfers can produce an excellent solution.

This collaging of images can be a spontaneous way of working, with the images juxtaposed in such a way that the original print is indecipherable. If the transfers are to be put one upon another it will be necessary to fire each layer before applying the next, as the covercoat and the adhesive gum must be burnt away in order to avoid the gases produced during firing from breaking through the next layer of transfer, fracturing the print and producing a frizzled appearance when taken from the kiln.

Transfers may be applied to bisquit, but the risk of faults is increased. It is more difficult to achieve a perfect contact between the print and the absorbent clay surface, particularly if the clay has anything other than a very smooth surface. Absorbency of the clay may be overcome by treating it with size which, when dry, will give a surface upon which the transfer may be placed more easily. Even so, the firing must be very slow (four hours to reach 200° C) in the initial stages, in order to vaporize the size without damaging the transferred image. Printing on to bisquit may be done more easily by some form of direct printing, or printing from a copper plate engraved with the desired image. The heated plate is inked up with a ceramic ink and, while still hot, the image is printed on to tissue paper. Before the ink is cold and hard the paper is placed upon the bisquited form, so that the ink is in direct contact with the ceramic. The paper is then softened with warm water so that the bond between the paper and ink is decreased, while that between the ink and ceramic surface is increased, and the image pressed home firmly. The paper is then removed to reveal the image upon the ceramic, or it may be left in place to fire away during the hardening-on firing, which is necessary in order to fire out of the ceramic ink the oil medium, which would otherwise prevent the glaze from establishing a smooth, unbroken layer during the application and subsequent firing of the glaze.

This type of underglaze printing was used to produce the Staffordshire blue printed wares. Victorian tile-makers also used it to produce tiles with an underglaze outline of the pattern, which could be coloured after the glaze firing by relatively unskilled painters, usually women, who had only to keep the various colours within the printed boundaries, rather than attend to the drawing of the whole design.

Transfer papers, whether for lithographs or silkscreen printing, should be stored in a room or cupboard where the atmospheric humidity can be kept constant. The paper comes with its top side coated with gum. As the paper is highly absorbent, only the back of it is in direct contact with the atmospheric moisture. If the paper should become damp it will curl, with the gummed surface on the inside. This can make printing very inconvenient, and, more important, it means that the paper may not be dimensionally constant. If an image is printed on such paper there is no guarantee that it will fit exactly the dimensions for which it has been designed. This is particularly evident with designs produced to fit up against the edges of tiles and plates. Similarly, if the paper is stored under controlled conditions there should not be any frustrating dimensional inaccuracies after printing. All this assumes that the ceramic pieces themselves will be produced with only limited variations of shape and size. Some types of printed imagery may be inappropriate for hand-made tiles, revealing otherwise unnoticed inaccuracies.

Lithographic printing produces a very light deposit of colour compared to those which are screen-printed. For this reason they may be more affected by the colour of the body, or, more usually, the glaze upon which they are fired. Colours which produce a dense, opaque layer when screen-printed may produce a semi-transparent deposit when lithographically printed. It would be most unusual to print with anything other than ceramic inks, prepared by a specialist supplier for the lithographic process.

Few ceramic studios would consider making lithographs for commercial products because it is so expensive. Even within the industry most of this type of work would be done by specialist firms who have the necessary skill to produce lithographic prints for use by ceramic manufacturers. The skills are those of the printer, and failing a technological miracle, no one would be able to carry out these processes without considerable experience.

Screen printing deposits a heavy layer of colour, the thickness of which may be felt if you run your finger across the surface of a fired piece of ceramic decorated in this way. The ink flows very slightly, after passing through the screen, thus avoiding the texture of the screen showing on the print. While it is possible to print non-commercially produced colours by silkscreen printing, the preparation must be thorough and accurate. The colour, whether it is under-glaze, in-glaze or on-glaze, glaze or oxide, must be ground to a fineness and consistency enabling it to pass through the screen without blocking it, or producing irregularities in the printed image. It is not usual to print anything other than prepared screen-printing colours for transfers. Direct screen printing allows much greater choice of medium, as the difficulties of holding the colour on the transfer do not arise.

Some individual artist-craftsmen maintain a capacity to produce their own screen-printed transfers, but this is usually only economic if the final product can command a price which justifies the time and effort involved. Others find printers who are prepared to undertake the making of transfers. Even if they have no previous experience, their skill as printers can often enable them to produce satisfactory results once they understand the process and have the right materials.

FIRING PRINTED DECORATION

It is essential that any kiln used for firing printed decoration is absolutely clean. It is best if one kiln, together with the appropriate furniture, is set aside specifically for this purpose. Any dust within the kiln atmosphere may be deposited upon the printed image, which will soften to some degree during the firing. These particles can produce a series of raised spots upon the finished article. Even though they may be invisible, they will be felt when the piece is handled, and will hold any dirt which finds its way on to the object when in use.

When transfers of any kind are fired the initial stages of the firing must be very slow, so that any medium used in the printing of the colour, or to fasten it on to the ceramic, can be driven off without disturbing the carefully printed image. During this stage of the firing the kiln should be well ventilated so that any vaporized material can escape from the kiln. Failure to do this may result in an oxygen-depleted atmosphere, which can have an adverse effect on many ceramic colours. Once all this medium has been burnt away the firing should proceed as fast as the kiln and the ceramic will safely allow. The final temperature will depend upon the colour or glaze which is in use.

Most colours used for on-glaze decoration should be fired to 750° C (1380° F) but some, especially those applied to bone china, may require a higher temperature. The supplier of the colours should specify the temperature range which will give the best results. Some colours fade with frequent firings, so if several colours are to be direct-printed and fired individually, those likely to fade should be fired last.

Where colours are to be fired higher than specified by the manufacturer – for instance, if they are to be fired into the glaze, but are supplied as on-glaze colours – experiments should be undertaken to determine the effects. In many cases the results will be quite acceptable, but in others the colour may fade or change dramatically.

Under-glaze colours require a hardening-on firing to drive off the medium before glazing and to attach the colour securely to the ceramic, so that it is not damaged during the glazing process.

In-glaze colours must be fired to the maturing temperature of the glaze so that they are completely integrated. On-glaze colours must contain some flux so that they will bond on to the glaze even though the latter may not soften much at the temperature to which the on-glaze colour is fired.

Cermifax

The Cermifax process is a development by Kodak which enables a glazed ceramic surface to be photosensitized. Upon exposure to light the surface becomes tacky, to a greater or lesser degree according to the amount of light falling on it. When dusted with a ceramic colour an image is produced with complete tonal accuracy, and without any of the halftone dots or lines which are normally associated with the printing of photographic imagery. It is exactly like a bromide print, except that the image can withstand much greater wear when fired. At this time there have been several attempts to establish a commercial method of using this technique, but so far none has proved economically viable for mass production. The technique is difficult to standardize, and while a success ratio of one in ten may be acceptable if only three or four very special and expensive items are required, it would be commercially disastrous to establish a production capacity on this basis.

The method of reproducing the photographic image requires the negative to be in almost direct contact with the sensitive ceramic surface. If the negative is taken farther away the exposure time required to affect the sensitized surface is very greatly increased, and a diffused image is produced which tends to reduce the value of the process.

Fumes given off during this process are noxious and special ventilation is necessary.

13 Design for infinite surfaces

This chapter deals not with the design of objects but with how a surface pattern may be designed. For the 'one-off' designed for a specific location there are no rules. Anything is possible, though not necessarily desirable. Whatever the type of application being considered, each will depend upon the imagination and inventiveness of the artist for the quality of surface treatment. In architectural environments the idiosyncratic can become an embarrassment if it depends upon its uniqueness rather than its inherent aesthetic quality. The one-off possibilities remain boundless, but designs for tiles intended for mass production, or to cover an unspecified surface area, must abide by specific rules if they are to work. In this sense the term 'infinite' does not mean 'going on for ever' but rather 'unspecified', i.e. the boundaries of the surface are undefined.

The rules regarding systems of repeating images are those of two-dimensional or planar geometry. Those who find even the terms daunting will, I hope, find reassurance in the description and illustrations. The rules are quite familiar, although you may not have considered all the variations which are possible.

We are surrounded by geometric patterns, particularly those of us who live in towns and cities, but the range is a fraction of those which could be used. To some extent this is due to the ease with which the simple patterns are absorbed, soon becoming the only types ever to be considered. Some geometric patterns are too complex for everyday use, and others may be drawn but do not lend themselves to fabrication if the angles or sides are too small for the material. These factors would be of considerable importance in tile manufacture, either by hand or automatic production. Patterns reached great geometric richness and variety in those countries which were predominantly Islamic in their faith. Islam discouraged the representation of any of God's creations, birds, animals and flowers, therefore the creative energy of artists and craftsmen was diverted into pattern-making. At the same time the Arab world had become the repository of Greek mathematics, including Euclidean geometry, during the Dark Ages in Europe.

Some shapes have proved more popular than others, and in the case of Islamic tiles there are certain classical combinations. Where these tiles are applied to the interior or exterior

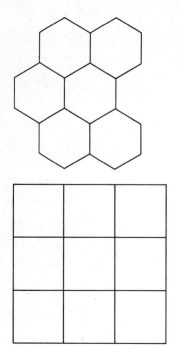

116 Regular tessellations.

117 Semi-regular tessellations.

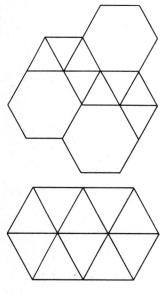

of a dome the mathematical precision involved in manipulating the tessellation so that it grows from the central point is an indication of the sophistication achieved by the tile-making workshops in the Near East.

TESSELLATIONS

The simplest type of pattern for an infinite surface is one that uses a single shape, which may be repeated, or tessellated, without any alteration, and will cover a surface without spaces between the shapes. Upon consideration of the simple geometry involved, it becomes evident that there are very few regular polygons which will tessellate. You can check the tessellating possibilities of any regular polygon (one with all its sides and angles equal), by determining the angle at one of the corners. If it is a factor of 360°, the number of degrees measured in one revolution around a single point, then the polygon will tessellate. Little experiment will be necessary to reveal that there are only three regular polygons which will tessellate. These are: an equilateral triangle, a square and a hexagon. A tessellation made up of any one of these polygons is known as a regular tessellation. Combinations of two or more of these regular polygons are called semi-regular tessellations, and there are only five possible combinations. Using a regular octagon (8-sided) and a regular dodecagon (12-sided) a further three semi-regular tessellations become possible. These three regular and eight semi-regular tessellations are the only ones possible on a flat surface, although there are an infinite number of variations on these eleven basic patterns.

On this foundation of regular and semi-regular tessellations are built all the tiling systems and patterns which can be repeated *ad infinitum*, whether they be purely abstract, geometric, figurative or floral patterns. Sometimes the underlying system may be obscured by the complexity and subtlety of the design. Nevertheless, if the pattern can be repeated to cover any surface, then the basis of the repeat must be one or other of these tessellating systems. The most commonly used are the repeating square and triangle, but the hexagonal tessellation is also frequently used.

The regular polygons may be modified in various ways to produce more interesting patterns, and irregular polygons may be regarded as related to, or variations of, the regular parent. A rectangle may be seen to be similar to a square in the way it behaves as a tessellating shape, but it will have two aspects: one with the shorter side as the base and one with the longer side as the base. In this case the tessellation remains the same but our view of it changes. All regular polygons which provide regular tessellations may be modified in this way with similar results. Groups of the tessellating shapes may be joined together to produce non-regular

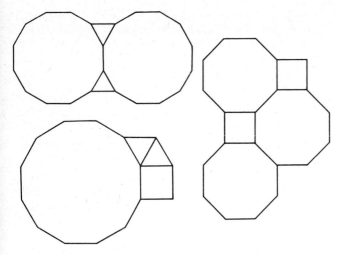

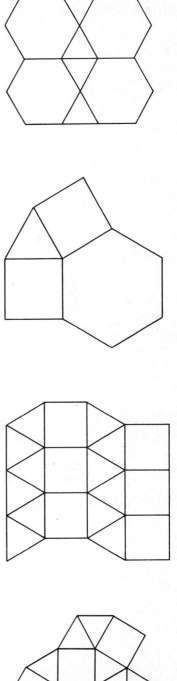

polygons. These shapes may be repeated to produce a further range of patterns.

Even without the addition of colour to these tessellations many varieties of each regular and semi-regular group are possible. If you consider the boundary of each shape you will see that it is possible to modify this edge, still allowing it to fit a similarly modified edge of the adjacent piece. Find the centre point of an edge and alter the line which connects the corner of the shape to the centre point. Trace this modification and turn the tracing paper over so that the traced line runs from the centre of the edge to the next corner. This will give a boundary which is symmetrical. You have in fact divided the edge into two parts, each part being the reflection of the other. If this edge, which we will call X, is then fitted to the next shape turned upside-down (through 180°) with a similarly modified edge X_1, there will be a perfect contact. The tessellating shape may have all its edges similarly modified, so that any edge will fit another. Alternatively, each edge may have a different modification and, provided that each edge is symmetrical about its centre, the pieces will fit together to provide a perfect tessellation. With the latter type it is essential that the modifications to the edges follow the same sequence on each piece when read in a clockwise direction, so that edge *a* is followed by edge *b* which in its turn is followed by edge *c* and so on. It is very easy, when designing on this basis, to mistake one edge for another, in which case the system will break down and will not create a surface capable of infinite extension. Of course, the danger is increased as the number of sides of the polygon increases.

Much has been written about this type of boundary-changing to develop tessellating systems. Those books which treat the technique as the basis of mathematical games are of particular interest to designers, as they approximate more closely to the creative process of the pattern designer. Varieties of modifications are shown in the illustrations,

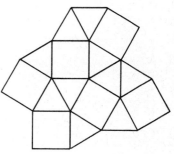

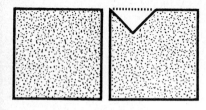

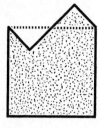

118 Boundary-changing for pattern-making. A piece is removed from one edge of a square tile, leaving at least half the edge intact. Replaced on the other end of that edge, it produces a side which is symmetrical around its midpoint, and the edge will interlock with that of an identical tile.

119 Three stages in developing a repeating pattern on a square module. *Top:* The pattern is drawn in the centre of the paper so that it touches all four midpoints of the edges. *Centre:* The paper is divided (irregularly, to disguise the repeat) across the midpoints of the edges. *Bottom:* The paper is relocated to allow the pattern to be continued with a guarantee of a perfect repeat. The edges with the same letter coding in the middle diagram must be brought together, *a* to *a*, *b* to *b*, etc.

and it will be obvious that variations on the basic tessellating systems need not be applied only as edge modifications, but may be the basis of printed patterns on regular tessellating tiles.

Another way of considering the surface modification is to join up the centres of each tessellating shape. Various secondary but linear patterns will be revealed. These linear patterns are called 'lattices', and are determined by the system of locating the basic tessellating shape. These lattices are the structure upon which various images may be applied on one tile, with the assurance that the pattern will run continuously across all the tiles when they are put together. The lattice of straight lines may be further modified within each tile and, provided the pattern contacts the edge of the tile at the point coincident with the original lattice, it will run across all the tiles, with each tile bearing a slightly different pattern. Such design solutions lie between the 'one-off' and mass production. They may not find much favour with large factories, but are suitable for small workshop production.

Another technique for developing tile patterns is best considered as a drawn design. Any configuration is drawn upon the centre of a square piece of paper. The paper is then cut, or torn, as in Fig. 119, and reassembled. The pattern is now on the outside of the paper and may be continued into the centre, so that it runs, uninterrupted, across the joins of the paper. When relocated in its original form the pattern will always repeat without interruption.

Where any of these continuous patterns are applied to ceramic tiles the grouting, which is an essential feature of the tile, should be stained to blend with the pattern, otherwise the boundaries of the tile may produce such a dominant lattice that the decorative pattern may be overwhelmed. In the case of geometric patterns, however, the grouting could produce an interesting counterpoint to the decoration.

None of these techniques will guarantee good designs. They are only the basic grammar from which rich patterns may be developed. While it is probable that the beginner may need to develop fluency within the confines of geometry, this should not restrict the development of more varied imagery. Dull geometric patterns can be very dull indeed.

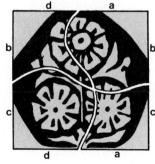

14 Glazes

A glaze is a form of glass which serves either to provide a more pleasing appearance to the ceramic object or to improve its performance by preventing penetration of water into the fired clay. Glass is a form of silicate and if it is to fit a ceramic material the glass should include some alumina. As both alumina and silica are the basic elements in clay it is natural that some form of clay material be included in the recipe of most glazes. To make the glaze fit the body at a convenient temperature, at least below that at which the body collapses, a flux is required in addition to the alumina and silica. Glazes are coloured by the presence or addition of metal oxides.

When glazing a piece of sculpture the cost of materials is not an essential factor in deciding on the type of glaze to use, although it would be foolhardy to disregard the cost altogether, as this information may be needed to establish the over-all cost of the project. In brick and tile production the price of glaze, engobe and firing, including the reduced density of packing which is inevitable in the glaze firing of these products, may be the deciding factor in whether or not the production is profitable. Too high a cost may make the product unsaleable against a cheaper competitor, but in many cases the high cost of glazing may be more than compensated for by the increased value of the brick or tile.

Bricks may be glazed to improve their visual quality but as this is so much more expensive than conventional finishes it has become limited to those installations where the bricks must be washed down in the interests of hygiene. These situations are limited as the modern practice would be to tile such surfaces. Before the improvement of tile production, glazed bricks constituted an economic structural ceramic with surface resistance to water penetration, so they may be found in public toilets, sewer installations and swimming pools, as well as in the interior wells of office blocks to eliminate cleaning and increase the level of reflected light. Tiles are glazed to produce decorative effects and to make them suitable for the type of environment previously requiring glazed bricks. When sculpture is glazed the reasons are usually aesthetic, but if the piece is to be placed outside, the glaze will tend to reduce penetration of the surface by water, thus resisting the tendency to crack in frosty weather.

GLAZED BRICKS

The commonest colours for glazed bricks are white, brown and black. Brown and black glazed bricks are usually the result of coating the brick with a clear transparent glaze, though in the case of the black some iron oxide may be added to the glaze to darken the final colour. On the whole, however, bricks can be glazed in any of the colours normally associated with other ceramics which are fired to the same temperature. As the body of the brick is likely to be quite dark it is normally necessary for the glaze to be opaque, so that it is not affected by the colour of the underlying clay. This is achieved by using one of the opacifiers, but as the cost would rule out this solution for mass production, it is more usual for the brick to be coated with a layer of white-burning clay, called an 'engobe'. The clay may then be transparent but unaffected by the colour of the brick. The engobe can be coloured by the addition of metal oxides or commercial stains, and functions as a kind of veneer on a clay surface. It is often cheaper to induce the finished colour in the engobe, which can then be coated with a clear transparent glaze. This would be the method of choice if bricks of several different colours were to be produced within the same works.

Many other colours are possible by using the processes normally associated with pottery finishes, but few of these would be suitable for situations subject to heavy wear, such as public floor surfaces. Even if the glaze were matt to avoid the hazard of slipping it can show signs of wear in quite a short time, and in due course would erode to reveal the colour of the brick.

Application of engobes and glazes to bricks

An engobe is normally applied to the appropriate surface of a brick while it is quite damp and certainly before it passes to the kiln. Mechanical fit can be a problem if the brick clay and the engobe have different rates of shrinkage from wet to dry. Similar problems may be encountered if the clay and engobe have different rates of thermal expansion, and the fault may not be evident until the brick has been fired. Problems of mechanical fit can be attributed to poor adhesion between the engobe and the brick. In order to achieve a good fit it is important to apply the engobe while the brick is damp but if this is not possible the engobe recipe must be modified to include some calcined clay, which will reduce shrinkage during drying. The thermal expansion of clays can vary considerably, and it may be necessary to substitute one engobe for another rather than go to the expense of modifying the first to make it fit the brick clay. As a last resort the thermal expansion of either the engobe or the clay may be increased by the addition of silica in the form of quartz or flint.

120 The Sanderson wallpaper factory in Chiswick, West London, designed in 1902 by C. F. A. Voysey in white glazed and black unglazed brick.

The glaze can be applied to the brick either before or after it is fired for the first time. The formulation of the glaze will vary according to when it is to be applied. If the glaze is to be put on to the green brick it must shrink at the same rate as the brick and engobe. To achieve this the glaze should contain a high proportion of clay in the recipe, which may be in the form of ball clay (high shrinkage) or china clay (lower shrinkage). If either of these is calcined the shrinkage from wet to dry will be further reduced. Montmorillonite clays, such as bentonite, are highly plastic ball clays and therefore have a very high shrinkage upon drying. Experiments must be carried out to determine the exact proportion of the various ingredients to give a satisfactory mechanical fit. The glaze will need some form of flux to act upon the clay component so that together they will melt at a convenient temperature and form a glaze with the desired characteristics. Lead has a thermal expansion similar to most clays which favour its use in glaze recipes. As the problem of soluble lead being passed from the glaze into food does not arise in glazed bricks, lead may still be used for this type of production. It should be in the insoluble form of lead silicate, either lead monosilicate, lead sesquisilicate or lead bisilicate, which are in the respective ratios of 1 to 1, 1.5 to 1 and 2 to 1 silica to lead. Lead starts to volatilize at 1140° C (2084° F), so above this temperature felspar is used as a flux. Borax in the form of a frit may be used throughout the temperature range. At lower temperatures a commercial frit, probably including some lead, may be used, and above 1140° C natural frits, such as colemanite, are suitable and relatively cheap. None of these glazes will show much evidence of colour apart from a slight yellow cast due to iron impurities in the ingredients.

Colour is produced by the inclusion of metallic oxides. In some cases these will give different results according to the atmosphere in the kiln during firing.

GLAZED TILES

Extruded tiles are treated in the same way as bricks because the methods of manufacture are very similar. This means that they are usually glazed while the tile is green and then fired once only. Wall tiles which are made from pressed dust are usually bisquit-fired and then decorated and glazed, if possible in one operation.

As tiles are bisquit-fired they may be glazed and decorated with the same material and techniques as those used in the manufacture of pottery. Glaze fit is carefully controlled because tiles which become crazed, or show some similar fault, may be rejected by the client, and if crazing occurs when the tiles have been installed their appearance will soon deteriorate. Crazing is caused by the body having a lower thermal expansion than the glaze, which allows the glaze

121 Spraying glaze on to tiles using a spray gun in an extraction booth.

at some stage in its working life to lie in a state of tension with the body; i.e. it aspires to cover a smaller area than the body on which it lies. As the body and the glaze are not made up of exactly the same elements, there is bound to be some difference, no matter how slight, in their thermal expansions. Even if they could be composed so that the thermal expansions are equal, crazing is likely to occur if the body should take up any moisture from the atmosphere of the immediate environment expanding as the moisture is absorbed. For example, moisture may penetrate the back of a porous tile if the surface to which it is bonded becomes damp, causing it to expand. The ideal state for the body and glaze is that the glaze should be in a slight state of compression so that, if the body should expand, this will improve the glaze–body fit. Like all ceramic materials the fired glaze has a much greater compressive than tensile strength. Thermal expansion of the body is determined by the materials included in its composition, and the temperature to which it is fired. The higher the body is fired the more free silica is converted to cristobalite. As this, together with the free silica, contracts considerably upon cooling, and at temperatures below the softening point of the glaze, the glaze will be thrown into compression as the body cools.

Crazing resistance is measured by placing the sample in an autoclave and subjecting it to steam under pressure. If the sample withstands 100 lb. per square inch (7 kg per sq. cm) for two hours without crazing it should not craze within its working life.

Thermal expansion factors of glaze materials:

material	chemical formula	factor ($\times 10^{-7}$)
alumina	Al_2O_3	16.7
barium oxide	BaO	14
boric oxide	B_2O_3	6.53
calcium oxide	CaO	16.3
magnesium oxide	MgO	4.5
potassium oxide	K_2O	39
lead oxide	PbO	10.6
silica	SiO_2	0.5
sodium oxide	Na_2O	41.6
zinc oxide	ZnO	7

From the above table it is clear that some materials have a very high thermal expansion. Those glazes which contain a high proportion of high thermal expansion material will tend to craze and those with very low thermal expansions may tend to peel. Peeling is the reverse of crazing, but because the compressive strength of glazes is high, peeling is much more rare than crazing. While both faults may have the same appearance, peeling is identified by overlapping of the separated pieces of glaze, and this can be felt if you run your finger across the glaze. Crazing and peeling can be cured by substituting for the offending material something with a more suitable thermal expansion. In theory this can be calculated to determine the precise amount of the substituted material, but in practice this modification of the glaze will have side effects which demand that the solution be found by trial and error in order to maintain the glaze qualities originally required. The common cure for crazing is to increase the amount of free silica in the body, or to increase the silica content in the glaze. When dissolved in the glaze, silica has a low thermal expansion, but such an addition will increase the maturing temperature of the glaze.

As most industrially produced tiles are either flat or have only slight surface modification the commonest use of glaze is to produce a decorative effect while maintaining the necessary technical qualities. The range of tile glazes is infinitely varied in tone, hue and texture. Glazes may be put one upon the other to produce reactions which cannot be reproduced in any other medium. These have been very popular in recent years but the qualities achieved have not been exploited in the most discriminating way. Such glazes are termed 'reactive glazes'. Others which have been popular are crystal glazes, speckle glazes and artistic glazes. These, together with the various colours, textures and methods of

122 Over-all view of the waterfall glazing machine. The glaze runs from the header tank (top right) through the wedge-shaped reservoir to the collecting trough and the bottom tank; from here it is pumped up to the header tank again. The tiles can be seen on a conveyor belt passing from left to right through the broad, uninterrupted sheet of glaze as it leaves the reservoir.

printing are the staple diet of the tile industry. They are characterized by producing a very active surface which, if applied to anything other than a flat tile, would overload the form of the object unless under the control of the most sensitive artist or designer.

GLAZING LARGE CERAMICS

The prime difficulty when glazing sculptural pieces, or any form which has a good deal of surface detail, is that the glaze tends to lie as a skin over the clay, obliterating the surface qualities of the form. To avoid this problem several types of glazing techniques may be tried, as unglazed clay may be unsatisfactory from a functional or aesthetic point of view.

The commonest method of glazing large-scale ceramics is spraying, using a compressor to provide the air pressure necessary to operate the spray gun. It is unlikely that this type of object will fit the conventional spray booth, so the

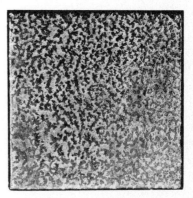

123 Rutile break glaze test, probably for glazes used in fireplace manufacture.

124 Glazing an insulator. The
insulator is held in the clamp and
rotated in the bath of glaze by means
of the handle at one end.

operation should be carried out away from any other people
and with good extraction or ventilation. The glaze being
sprayed may or may not include harmful ingredients, but
every effort must be made to ensure that no atomized glaze
is inhaled or finds its way on to everyday clothing, where it
may, sooner or later, be transported outside the glazing area.
If ventilation is not efficient the room will be filled with a
mist of glaze particles, which produce a fine film of glaze
upon every surface in the room. This should be removed,
once spraying has ceased, and may be re-used if uncontami-
nated. It is vital during spraying on this scale that you wear
appropriate protective clothing – Terylene or similar over-
alls, a respirator, goggles, gloves and a close-fitting hat.
All these should be removed and cleaned when you have
completed the glazing process.

If spraying facilities are not available, and you need to
apply glaze to a large form, it is possible to do so by painting
with a large soft brush (a glaze mop). The glaze should be
applied very generously, paying particular attention to any
part of the form which is recessed and upon which the
glaze does not naturally fall. Painted glaze tends to be very
uneven unless you have had considerable experience, therefore it is common practice to rub the dry glaze surface with a
coarse fettling material, such as wire wool or scrim. This is
akin to sandpapering a wood surface and has the same
smoothing effect. You should be careful not to rub off too
much glaze in any one area. If the glaze has run and dribbled
down the form, these runs will stand proud of the surface

and should be levelled by paring away with either a razor blade or a sharp fettling knife.

Some glazes do not lend themselves very easily to being painted on to a clay form, not that they present any physical difficulties but because they reveal characteristics of the method of application which can mar the quality of the form being glazed. Transparent or translucent coloured glazes are typical of those that present this problem. They give excellent finished surfaces with the added advantage of emphasizing minor modelling characteristics, because where the glaze is thicker the finished colour will be more pronounced. If the glaze has run, either during application or firing, this run will be stronger in colour than the surrounding glaze. It may be that you can use even this characteristic to good effect, but if not you should check for an even application of the glaze and take any remedial action which is necessary.

Some large insulators are glazed by dipping, and if the work has the necessary characteristics you can use the following method. The form must be hollow and very strong, with holes in the centre of the base and top. It should also be symmetrical in weight distribution around the central axis. The technique is to clamp an axle on to the form when it is horizontal, having provided a soft bed which will support the form until the axle is clamped in place. The piece is then lifted on two end cradles so that the clay form may be rotated by turning the axle. A bath of glaze slightly larger than the piece is prepared and placed

125 When the insulator has been glazed all over it is removed from the bath but the rotation is continued until the glaze has dried and there is no risk of runs.

beneath the clay form. The form is then lowered so that it is partially immersed in the glaze. Turning the form slowly will produce a skin of glaze over the entire surface. It is important to ensure that any undercut areas are immersed in the bath, unless these are to be unglazed. The glaze soon dries and any runs can be fettled off.

Vapour glazing

Vapour glazing with sodium chloride may be used where these facilities are available. The kiln must be reserved for salt glazing as, once used for this process, it is unsuitable for any other.

Whatever type of glaze and whatever method of application is decided upon they should be tested before attempting to glaze a large object. Too much haste at this stage may ruin the work of months. If any mistakes occur it will be necessary to refire the piece, which is both expensive and hazardous. It is bad enough firing the work once to perhaps 1200° C (2200° F): to do this more than is necessary is to court disaster.

When applying glaze by means of an airograph it is very easy to miss the underside of some surfaces and although these may appear to have been glazed by the spray of the airograph the fired result will show only a slight deposit of glaze. It is best if these areas which are difficult to reach are glazed first and any areas that are glazed accidentally may be wiped off with a clean sponge before commencing to glaze the rest of the form. This cleaning off is necessary to avoid overglazing those parts which cannot be avoided during the normal process of spraying the glaze on to the form.

You must be careful to avoid heavy deposits of glaze at the base of the piece as this may tend to run, and fasten the work to the kiln shelf upon which it sits during the firing. Such heavy deposits can be fettled off with a sharp knife or razor blade before firing. Alternatively, these areas may be coated with wax or wax emulsion to ensure that no glaze adheres to the bottom half-inch of the clay.

When glazing large clay pieces which have not first been bisquited you must be careful not to introduce too much water on to the clay as this might weaken it and cause it to collapse or crack.

SALT GLAZING

Salt glazing relies upon the clay containing a high proportion of silica, sometimes in the form of sand. When the clay is fired to at least 1180° C (2156° F), salt is thrown into the kiln chamber where it volatilizes and penetrates the outer surface of the clay. The salt must be damp to assist the vapour to penetrate the clay, which should be beginning to vitrify.

Five per cent of water may be added to dry salt. Then it combines with the silica to form a thin glaze. The saltings may be repeated several times and after each introduction you must draw a trial piece, previously set in such a way that it will receive a fair share of the salt vapour, then be withdrawn, cooled, and inspected to determine the progress of the glazing. You may need to set a number of these draw trials if several saltings are to occur. If the salt is introduced at too low a temperature the glaze will penetrate too deeply into the clay and give a thin surface glaze. The colour of the glaze will be determined by the amount and type of metal oxide present in the clay itself or in any engobe which has been applied. Any ceramic surface within the kiln will be liable to attack by the salt, so furniture used to stack the kiln must be coated with alumina if it is not to be glazed; also the amount of salt required to glaze the form will be greater than expected if the interior of the kiln is absorbing some of it.

In practice, once a kiln has been used for salt glazing it is kept for that purpose and no other. After several firings the amount of salt required to produce a good glaze will be less than on previous firings because the glaze is no longer taken up by the kiln brickwork. The life of a salt-glazing kiln may not be more than five years as the constant attack upon the brickwork will gradually reduce its mechanical strength. The surfaces of the object may not be evenly glazed as the salt vapour is carried by the kiln atmosphere; the interior of pipe and bowl forms may not receive any glaze at all. In the manufacture of sewer pipes, where this technique is still in use, it is common practice to glaze the interior of the pieces with a conventional glaze before firing. Some manufacturers glaze their products with conventional glaze and then subject them to salt firing to produce the distinctive 'orange peel' effect with a lower loss rate.

Salt glazing has the advantage of being developed upon the surface of the clay, therefore little of the original surface quality is lost and the glaze does not craze. On the other hand the kiln cannot be used for any other type of firing because even if no salt is added to the chamber, that which has been taken up by the kiln brickwork in previous firings will be released during the firing and produce salt-glazed surfaces upon the objects being fired. The chimney of the kiln should be kept in good condition to ensure that the toxic gases produced by the salt glazing (hydrochloric acid fumes and small amounts of chlorine vapour) are drawn well away from the kiln and the surrounding area. You may be fortunate enough to find a manufacturer of salt-glaze ware in your area who will allow you to fire your pieces in his kiln. As the quality of salt glazing can vary from firing to firing, and there may be a good deal of variation within the chamber, make sure that your piece of work is fired in the most suitable place in the kiln, and that the kiln is carefully fired.

During firing the kiln should be well oxidized to ensure that the clay is fired properly and without reduction faults. When the salting process is under way the damper of the flue should be closed to hold back the vapour so that it will concentrate its effect upon the clay in the kiln and not simply rush out into the atmosphere. This may produce reduction qualities in the glaze if maintained for a long time.

Bricks may be salt-glazed but the open packing needed to glaze the brick all over is normally avoided so that only those surfaces of the brick which will be exposed when installed are glazed.

Soda ash (sodium carbonate) is generally considered too soluble to be used without first fritting it. However, if it is applied to the surface of an unfired clay form it will penetrate the clay and then rise to the surface with the water in the clay as the drying process proceeds. It is deposited on the surface as a light crust and when fired fluxes the silica and alumina to form a glaze. The temperature range is wide, the deepest and most permanent glazes being those fired above $1100°$ C ($2012°$ F). The soda ash may be applied to bisquit clay and the resultant glaze will be thinner because less water is involved in the prefiring stage. Some of the water which penetrates the bisquit will evaporate and leave the sodium within the clay rather than on the surface. This technique lends itself to objects which demand an unregulated surface, and will give unusual colour and texture effects when used with commercial colours.

The soda ash, prepared as a slush of water and soda ash, is painted upon the clay and several layers may be applied to deepen the thickness of the glaze. The solution is highly caustic and if you have any small cuts on your hands you should protect them with rubber gloves.

A gentler approach is to dissolve the soda ash completely in water until the solution is only just saturated (hot water will accept a higher concentration). Once the solution, or slush, has started to crystallize it is very difficult to dissolve it again. Make up only the quantity that you immediately require. When the crust has developed on the clay it is quite strong and the object may be packed in the kiln with no more than the normal care. If the crust is thick it may shell off during the initial stages of the firing. It seems probable that this is caused by the driving out of water from the damper parts of the clay. As it forces its way through the outer surfaces the water dislocates the soda ash crystals. Kiln shelves should be coated with sand to prevent any of these droppers from fusing on to them during firing. Similarly, any other pieces being fired in the kiln should be protected by walling in the offensive piece with kiln shelves coated with bat wash.

If the firing is in the stoneware range the glaze will be brighter and glossier than those fired at lower temperatures. If it is fired below $1140°$ C ($2080°$ F) the soda glaze may prove soluble and gradually change quality through atmospheric attack. The addition of petalite may improve the hardness and quality of the glaze.

CRYSTAL GLAZES

These are glazes which develop a crystalline structure during the cooling phase of the glaze firing cycle. In some cases this will be enough to produce a matt glaze, but in more extreme cases the glaze will be smooth and shiny with crystal patterns developed in the glaze. The latter can be very dramatic but have tended to become debased by indecorous applications. To produce large crystals there should be very little alumina, therefore the recipe should contain only very small amounts of clay materials and very little felspar as all these contain too high a proportion of alumina. A side effect of reducing, or eliminating, the alumina is that the viscosity, and hence the maturing range of the glaze, is reduced so that it will run as soon as it melts.

The crystals in this type of glaze are either zinc silicate (willemite) or calcium silicate (wollastonite); they are flat, and so do not project through the glaze. The silica content of the glazes should be lower than normal and the additional glass-making ingredient is borax in the form of a natural or commercial frit.

Aventurine glazes

These are crystal glazes produced with an overloading (up to 20 per cent) of iron and chrome oxides or iron chromate. The crystals may form all over the surface of the glaze or they may be in large patches separated by areas of shiny opaque glaze.

When coloured with iron or metal oxides the colour may be isolated into patches around the crystals rather than dispersed throughout the glaze. Some crystals tend to grow on the more horizontal surface of the piece, while others seem to favour those parts where the glaze flows more, i.e. on the more vertical planes. Even more frustrating is the tendency for glazes applied to the inside of enclosed forms to produce better crystal effects than on the exterior. No doubt this is due to the heat loss on the interior being much slower than on the exterior.

Being crystal glazes, aventurines are highly liquid when melted, owing to the complete, or almost complete, absence of alumina. If there is a danger of the glaze running off the piece altogether it should be set in the kiln on an unfired or soft-fired piece of clay which can collect the glaze and be ground off when cool.

A technique favoured by some manufacturers is to coat the ceramic object with a more stable glaze which is compatible with crystal glazes, but including the normal proportions of alumina and silica so that it does not run when fired. The crystal glaze may then be applied as a second glaze and directed to produce a more controlled location of crystals in the finished piece. Nevertheless, it is quite common to find crystal-glazed pottery with bases showing evidence of substantial grinding to remove run glaze.

A glaze which fires between 950° C and 1050° C (1740°–1920° F) and produces calcium silicate crystals is favoured for the base glaze for producing aventurine glaze effects. The glaze is unable to hold all the iron in solution as it cools and this results in the excess material crystallizing out.

Of the true crystalline glazes, the zinc silicate type is the commonest. The recipe will often contain some titanium oxide to act as seed and encourage the growth of other crystals. Rutile, which is a form of titanium oxide with iron present as in impurity, may be used in lead glazes, fired to 1050°–1080° C (1920°–1976° F), to produce crystals of lead titanate. Titanium tends to crystallize out of any glaze during the cooling cycle, therefore the glaze composition is much less critical and the size of the crystals may be modified by the use of coarse- or fine-ground rutile. The quantity added to any glaze may range from 5 to 12 per cent but in all cases the iron impurity tends to give a brown tinge to the 'break-up' of colour.

With the possible exception of 'rutile break' glazes the firing cycle for crystal glazes is crucial. The firing may proceed as normal to the maturing point of the glaze, but from that moment on the cooling must be as slow as possible, and few kilns will produce the necessary slow cooling if simply switched off. The kiln should therefore be 'fired down', i.e. heat should be reintroduced from time to time so that the temperature reduces very slowly. Experiment will be necessary to achieve the required degree of control.

Metal oxides may be introduced to all crystalline glazes to tint the crystals in various colours.

MATT GLAZES

These are produced by inducing small crystals in the glaze. Commercial matting agents are available in either 50/50 mixes of calcined china clay and zinc oxide, or equal parts of zinc, tin oxide and titanium oxide. The former tends to produce glazes with a very low thermal expansion which may crack a piece which is thinly made, and may promote crawling. The latter produces a matt vellum and has a glazed surface favoured for commercial tiles. Cooling too fast may reduce or even eliminate the matt quality.

Other matt effects may be induced by overloading the glaze with some refractory material, but these tend to produce an underfired rather than a matt appearance.

LUSTRE GLAZES

The term 'lustre glaze' describes a surface which reflects various light waves and produces diffraction effects; this results from the presence of metal salts fired under reduction conditions.

On-glaze lustre is commercially the most used of the two types of lustre glazing, as the metal salts can be combined with resin. The resinate is dissolved in an oil, lavender oil being the commonest, and the mixture is self-reducing, so that it may be fired in any kiln without the necessity of modifying the atmosphere. This technique produces standard results, and a wide range of colours, as well as gold and platinum, are available.

Originally the on-glaze lustres had to be fired in a reduction atmosphere and in many ways this technique produces richer qualities than the commercial variety. Glazes with a high lead content are considered to be the most suitable for reduction-fired on-glaze lustre, although good results have been achieved with some alkaline glazes.

The metals are applied as a mixture of the metal nitrate or carbonate, together with 70 to 80 per cent refractory clay (red ochre). This mixture is applied very thickly in the form of a paste and then fired in a reduction firing, after which the clay is removed by washing to reveal the lustre effect. In both types of on-glaze lustre the lustre firing is usually about 750° C (1380° F) but with the non-commercial lustre the possible range is much wider – 600° C to 800° C (1110°–1470° F) depending on the metal being fired.

In-glaze lustre is more durable than on-glaze because it is defended by the glaze. Between 2 and 15 per cent of metal salt, carbonate or nitrate, is added to the glaze which is normally a low-fired (1000°–1050° C, 1830°–1920° F) lead oxide, or lead borosilicate, glaze. Highly alkaline glazes have been successful in producing softer colours. The lead glaze may include about 10 or 15 per cent of tin oxide, together with some zinc oxide (2–8 per cent). It is normally fired to mature the glaze and then reheated to about 650° C to 700° C (1200°–1290° F) and reduced for about twenty or thirty minutes. Lead glazes may be blackened by too much reduction, and if the colour reaction is good but the glaze is affected in this way it will be necessary to reduce the amount of lead in the basic glaze. There is no reason why the reduction should not be effected as the glaze is cooling rather than refiring, but it is usual to check that the glaze has fused properly before attempting to reduce the surface.

In any type of reduction firing for lustre effects, the point at which the reduction is started, the degree of reduction and the length of time during which the reduction is sustained will be critical in determining the final quality. With each kiln it will be necessary to experiment with different times and weights of reduction to achieve the best results.

ENAMELS

Enamels are low-fired glazes normally used as on-glaze colours. A wide range of colours is available commercially and they are fired in neutral or oxidizing atmospheres up to 750° C or 800° C (1290°–1470° F) according to the manufacturers' specifications, and the type of ware to which they are applied. They may be fired above these temperatures to produce glaze effects, or be over-fired so that they 'drop' into the glaze very slightly. In both these instances the colour produced may not be that which is specified by the manufacturers.

Glazes may be modified after firing by sand blasting or acid treatment. This is done from time to time if the glaze comes from the kiln with a less matt glaze than that desired. Acid is used to etch the glazed surface but should be handled only by those experienced in its use and in the necessary safety precautions.

ELEMENTS USED IN GLAZES

Glass-makers

Silica and boron are the commonest elements which will melt to produce a glass. They are plentiful, and although glazes can be made without the inclusion of any boron the latter is never used without some silica. Boron is available only in soluble form and must be fritted to allow for its easy use in glaze composition.

Glass-modifiers

Alumina is the main modifier of the glass-forming materials. It serves to increase the viscosity of the glaze and thus prevent its running off vertical or near-vertical surfaces. Rather refractory, alumina tends to increase the maturing temperature of the glaze, and less will be present in low-temperature glazes (up to 1150° C, 2100° F) than in high-temperature glazes (above 1200° C, 2190° F). It also helps to extend the maturing range of the glaze so that it is possible to fire within a reasonable range of temperatures. This is necessary given the inevitable slight variations in temperature which occur in normal production kilns, and with the temperature-measuring equipment that is available. The hardness of glazes is improved by the addition of alumina, so that it becomes less liable to damage. It also inhibits recrystallization of the rest of the ingredients in the glaze during cooling. For this reason alumina is absent, or retained in very small quantities, in recipes for crystal glazes.

Fluxes

Most other elements included in glaze recipes are intended to make the glaze melt at the desired temperature and to produce the matt or glossy surface required. The following alphabetical list of elements explains the contribution of each element, but these are introduced into the glaze by means of the materials listed overleaf. As many of the materials are naturally available in combination with others, or silica and/or alumina, it is much cheaper to use them rather than refined chemical elements.

Barium, toxic. A secondary flux in most glazes, but very often used to induce mattness except in the presence of glazes containing boron. COLOUR: Tends to produce blue-greens when copper oxide is present.

Boron. Although considered as a glass-making element, boron also acts as a flux in all glazes. In combination with lead, boron produces hard-wearing, trouble-free glazes with acceptable thermal expansion, which makes it relatively craze-resistant. COLOUR: It tends to heighten colour but the tendency to boil during melting may produce mottling in some glazes.

Calcium. A secondary flux at low temperatures but a primary flux at high temperatures. The resultant glaze is hard-wearing. COLOUR: Assists in the development of reduction-fired iron-bearing glazes.

Lead, toxic. A low-temperature flux which starts to volatilize above $1170°$ C $(2140°$ F). Lead-fluxed glazes are relatively trouble-free, with a lustrous type of surface and good craze-resistance. As lead is toxic it is normally used in a fritted form, and even then is suspect when used with certain colours on tableware. COLOUR: Favours apple-greens from copper but this is not suitable for tableware as the copper tends to increase the amount of lead released from the glaze.

Lithium. A strong alkaline flux which may produce glazes with less tendency to craze than those containing potassium or sodium. COLOUR: Produces alkaline responses in coloured glazes.

Magnesium. A high-temperature flux, but tends to encourage crawling of the glaze and is often used with at least one other flux. COLOUR: Tends to produce red or purple from cobalt.

Potassium. A strong alkaline flux at all temperatures but soluble, and must therefore be fritted for use in glazes. High thermal expansion may produce crazing if not compensated for. COLOUR: Favours blue-purple from manganese but all other colours are typically alkaline.

Sodium. A strong flux at all temperatures but is used in a fritted form as it is soluble otherwise. Glazes are soft and apt to craze. COLOUR: Promotes brilliant colours, especially turquoise from copper and red-purple from manganese.

Zinc. Used to produce matt glazes as the zinc crystallizes out of most glazes when cooled, but in excess will encourage crawling of the glaze. Important ingredient in crystalline glazes. COLOUR: Chrome produces brown rather than green, and copper tends towards turquoise.

Alkaline frit. A mixture of alkaline material such as sodium or potassium together with silica. The ingredients are melted together to form various silicates which are insoluble and therefore easier to handle in glaze-making. Commercial frits are available for use at various temperatures.

Aluminium hydrate. Occasionally used to introduce the necessary alumina into a glaze recipe, but more often as a bat wash.

Ash. The burnt remains of any natural vegetable matter will contain some silica together with alkaline elements and perhaps some metallic oxides. If the ash is properly prepared it may be used to form the basis of a glaze.

Ball clay. Contributes alumina and silica to glazes and has the advantage of assisting in keeping wet glazes in suspension while in store.

Barium carbonate. Toxic; a flux in some glazes if the quantity does not exceed 10 per cent. Above this amount it tends to produce mattness in the glaze, except when boric oxide is also present.

Bentonite. A ball clay made up of very small particles and used as a suspension agent in glazes (2–3 per cent).

Bismuth. In the oxide form, takes the place of lead in some glazes but more commonly used in small workshops in the form of bismuth subnitrate to produce mother-of-pearl (nacreous) lustre at low temperatures.

Borax frit. A mixture of borax and silica melted together to render the borax insoluble. Various commercial borax frits are available and some may include other materials.

Calcium borate (colemanite, borocalcite). A natural frit of boric and calcium oxides. An active flux in low-temperature glazes (5–40 per cent). A secondary flux in high-temperature glazes (5–15 per cent).

Calcium carbonate (chalk, whiting). A flux in high-temperature glazes up to 25 per cent, but more than this will tend to matt the glaze.

Calcium magnesium carbonate (dolomite). A secondary flux in high-temperature glazes (5–25 per cent).

Calcium phosphate (bone ash). A secondary flux in high-temperature glazes (5–10 per cent).

China clay (kaolin). A white-firing form of non-plastic clay used to introduce alumina and silica into glaze (5–25 per cent).

China stone (Cornwall stone). Similar to other felspars but contains a higher proportion of silica. A high-temperature flux suitable for use as a secondary flux at low temperatures.

Felspar. A natural mineral containing sodium, potassium, silica and alumina in various proportions. A low-temperature secondary flux (10 per cent) and a primary high-temperature flux (50–80 per cent).

Flint. A form of pure silica, used to supply the silica not provided by other materials in the glaze recipe.

Lead bisilicate. Toxic; a frit which acts as a flux in low-temperature glazes. Contains lead and silica in the proportions of 1:2. Commercially available.

Lead frit. Toxic; a combination of lead and silica together with other ingredients such as borax which, when fritted together, will act as a flux in low-temperature glazes. Commercially available.

Lead monosilicate. Toxic; as lead bisilicate, but the lead and silica are combined in the ratio of 1:1.

Lead sesquisilicate. Toxic; as lead bisilicate but the lead and silica are combined in the ratio of $1:1\frac{1}{2}$.

Lepidolite. A natural frit of lithium, silica and alumina.

Lithium carbonate. If added to glazes ($1-1\frac{1}{2}$ per cent), improves the glossiness and hardness of the glaze. Useful in crystal glazes as an aid to crystallization if more than 1 per cent is present.

Magnesium carbonate. A high-temperature flux up to 10 per cent. Tends to produce a buttery type of glaze surface but more than 10 per cent makes for mattness.

Magnesium silicate. A secondary flux at all temperatures.

Nepheline syenite. A very fusible felspar. 10 per cent has the same melting properties as 15 per cent of other felspar.

Petalite. A natural mineral combining lithium, alumina and a high proportion of silica.

Potassium carbonate. Used as a source of potassium in glazes but usually added in the form of a frit as it is soluble in water and deliquescent.

Quartz. A natural source of pure silica although slightly less reactive in glazes than flint.

Sodium carbonate (soda ash). A soluble form of sodium, normally used in fritted form.

Sodium chloride (common salt). A form of sodium used in salt glazing.

Spodumene. A natural frit of lithium, alumina and silica.

Talc. A natural combination of magnesium and silica.

Volcanic ash. A fusible rock used in amounts from 10 to 50 per cent.

Zinc oxide. Used in small amounts (5 per cent) in high-temperature glazes as a flux and in all glazes to produce mattness (15 per cent), often in combination with other materials.

Water is used to suspend the glazes, to make application easier. Most tap water will contain soluble minerals (particularly alkalis) which may either affect the physical condition of the glaze suspension or be passed into the glaze and affect the ware. In most cases this is not noticed but from time to time these ingredients in the water may reach such proportions that the glaze is substantially modified, usually for the worse. Such water may have some effect upon the clays or slip being used. Most water companies will provide an analysis, if this is wanted in order to trace some aberration in either glaze results or making procedures.

Colour is introduced into glazes by means of metal oxides or carbonates which will affect the glaze in various ways depending upon the formulation, firing temperature and atmospheric conditions within the kiln during firing. The following list of materials includes information on how these conditions affect the colouring agent.

Antimonate of lead. Toxic; 5–10 per cent will produce yellow in lead glazes.

Antimony oxide. Toxic; 7–17 per cent will produce white, although in lead glazes some yellow will occur as lead antimonate is formed.

Cadmium. Toxic; together with selenium will produce a range of colours from yellow to red, normally in an alkaline glaze. Most easily used in the form of a commercial glaze. The colour is destroyed by the presence of copper oxide which turns it brown or black and eventually green if the concentration is high enough.

Chrome oxide. Toxic; $\frac{1}{2}$–3 per cent will produce green in engobes and glazes, which becomes brown in the presence of zinc oxide. It can be volatile in some conditions and will produce pink in the presence of tin oxide.

Cobalt carbonate. Toxic; produces blue and may be preferred to the oxide as a more even distribution of colour can be achieved.

Cobalt oxide. Toxic; $\frac{1}{4}$–2 per cent produces blue in all glazes and engobes.

Copper oxide. Toxic; 1–5 per cent produces green in most glazes, clays and engobes. In alkaline glazes the colour tends towards turquoise depending upon the amount of alumina, and towards blue in the presence of barium. Under reduction conditions the result may be brownish red or purple in the presence of tin (3 per cent) and boron.

Copper carbonate. Toxic; finer grains than copper oxide.

Crocus martis. An impure form of iron giving mottled brown and yellow in stoneware glazes.

Ilmenite. A mineral containing iron and titanium which produces brown speckles in glazes. Available in various grades according to grain size.

Iron oxide (red, yellow, purple and magnetic iron). In lead glazes 1–4 per cent iron will produce yellow to amber; 7 per cent gives brown. The colour is brown in alkaline glazes. In combination with tin it tends to produce speckles at low temperatures. In stoneware glazes containing whiting and felspar $\frac{1}{2}$–2 per cent will give various shades of green when fired under reduction conditions. If tin is present the colour may be grey, and if calcium phosphate 3–7 per cent then the colour may be blue-green. 5–8 per cent will produce brown to red and 10 per cent will give black where thickly applied and red-brown where thin. Glazes containing more than 10 per cent may give red crystalline effects when fired in reduction, particularly over red-firing clay. Magnetic iron is favoured for aventurine glazes or speckled effects.

Manganese dioxide. 2–10 per cent gives brown in lead glazes and tends to purple in alkaline.

Nickel oxide. Toxic; 1–3 per cent produces brown or green but is more commonly used to modify other colours. At high temperatures, if crystals are produced and zinc is present, the colour may be yellow, purple or blue.

Potassium bichromate. Toxic; a combination of chrome and potassium which gives red-orange in low-temperature high-lead glazes.

Rutile. A mineral containing iron and titanium. Though containing less iron than ilmenite, it is similar in the effect that it has in glazes.

Selenium. See cadmium.

Tin. Up to 8 per cent will opacify most glazes to give white. In excess, it can cause crawling. It is not melted by the glaze but suspended in it.

Titanium oxide. 5–10 per cent will opacify and sometimes matt most glazes but the colour is more cream than that produced by tin as impurities are usually present. It is also used to seed crystalline glazes.

Uranium (depleted uranium oxide). Only occasionally used as it is now considered unsafe. Produces yellow or red in low-fired lead glazes. Grey results if the glaze is reduced.

Vanadium oxide. Toxic; 5–10 per cent gives yellow, usually in combination with tin oxide.

Zirconium oxide. Up to 15 per cent will opacify glazes to give white and it is used with chrome to give opaque greens, which would be pink in the presence of tin oxide.

Some metal oxides act as fluxes in glazes and although this may be disregarded in many instances, if the total amount of metal oxide exceeds 10 per cent there is a risk that the glaze will become more fluid, and tend to run on vertical or near-vertical surfaces.

Oxides may be combined to good effect either in the glaze or as a colour to be painted, printed or sprayed under or over the glaze. Commercial colours are either fritted forms of combinations of oxides – e.g. enamels for on-glaze decoration – or compounds of various oxides and stabilizers such as china clay. Coloured glazes with various degrees of mattness may be located one upon the other, and even those which tend to crawl or craze may be overglazed with other, more fluid, glazes to create decorative effects. Lead glazes may be set upon alkaline glazes which fire to the same temperature, with the almost certain result of producing a chemical reaction which may be dramatized by the presence of colouring and contrasting oxides in the glazes. Colours which attack other colours may be applied in small areas only to reveal both effects.

GLAZE PREPARATION AND TESTING

The materials for the glaze recipe should first be collected together so that there is no delay in finding the next ingredient when in the midst of compounding the glaze. The materials are weighed out according to the recipe and mixed with enough water to produce a smooth paste. This paste is sieved, normally through a 150- or 200-mesh sieve. The glaze-water suspension is then brought to the appropriate pint weight for the clay which is to be glazed and applied in the most suitable way.

Thick glazes can be easily thinned by the addition of more water but those which are too thin or watery may need to be boiled to drive off some of the water or allowed to settle overnight and the excess water decanted.

Glaze testing

Any material or group of materials may be tested in combination with another by combining them in the following system. Each combination should be fired upon a separate ceramic piece, either a tile or bowl. The latter has the advantage that if the resultant glaze is too liquid for the temperature to which it is fired, then it will not flow on to other trials or any part of the kiln.

Material A	10	20	30	40	50	60	70	80	90	%
Material B	90	80	70	60	50	40	30	20	10	%

If the most suitable result appears to lie between any one of these combinations the trials should be repeated within the range which looks most promising. For instance, if the best result seems likely to lie between 40%A + 60%B and 50%A + 50%B the trials are carried out:

Material A	41	42	43	44	45	46	47	48	49	%
Material B	59	58	57	56	55	54	53	52	51	%

This technique can be used for determining colour responses as well as glaze qualities.

It is most important, when testing glazes, to ensure that the firing conditions and cycle resemble as closely as possible the actual conditions under which the production pieces will be fired. Rapid firing in small test kilns may give results unobtainable in larger and slower-firing production kilns.

All glaze tests must be carefully recorded, so that all results can be reproduced (or, in the case of poor results, avoided).

RECIPES (The recipes below are expressed in percentages.)

Vitrifying engobes

The formulation of vitrifying engobes will depend upon the condition of the clay to which the engobe is to be applied.

	1	2	3	4	5	6
clay	100	40	25		25	
china clay		30	25	25	25	25
calcined china clay			25	25	25	25
borax frit				25		
flint		30	25	25	25	25
felspar						25
maximum firing temp.				1150° C		1250° C

1 = for leather-hard clay
2 = for dry clay
3 = for dry clay
4 = for bisquited earthenware
5 = for dry stoneware clay
6 = for bisquited stoneware clay

Earthenware glazes fired at 1050° C (1920° F)

	1	2	4	5	6
alkaline frit	75	70			70
whiting	3				
lead frit			63	83	
felspar	15				
flint	7		3		
zinc oxide		30			
cornish stone			26	17	15
barium carbonate			8		
china clay					15

Nos. 1, 4, 5 and 6 are clear, shiny glazes; No. 2 is very matt and tends to crawl when thickly applied but is useful when applied over or under other glazes.

Stoneware glazes fired at 1250°–1280° C (2282°–2335° F)

	1	2	3	4
felspar	70	40	20	78
china clay	13	10		2
whiting	12½	20	26	6
flint	4½	30	40	14
dolomite			15	
zinc oxide			3	
rutile			5	

Nos. 1, 2 and 4 should mature at 1280° C (2335° F) to provide clear smooth glazes but No. 3 will fire matt.

Crystalline glazes

	1	2
whiting	12	2½
barium carbonate	6	
borax frit	16	75
potash felspar	5	5
china clay	4	2½
flint	24	
zinc oxide	33	15

No. 1 should be fired to 1250° C (2282° F) and No. 2 to 1060° C (1940° F). Neither will be coloured, and 2–5 per cent titanium may be added, to act as seed crystals.

15 Installation

Any architectural commissions, whether they are for decorative bricks, wall panels, sculpture or tile facings, carry certain obligations to the client and to the public. I have no intention of trying to describe the intricacies of civil and criminal law, but it is essential to understand that advice should be sought from a solicitor or from a designer/ artist association, such as the Society of Industrial Artists in the United Kingdom or the Industrial Designers Society of America, before signing contracts or accepting any commitment which may increase your legal liability.

It is common practice to carry out work on the basis that responsibility for carriage and installation rests with the client, your liability being limited to faulty material and poor workmanship in the ceramic object only. Also, if you subcontract the work of installation you may be liable to the client for any faults which arise. Specifications for loadings, foundations for sites to support large sculptures, adhesives and methods of installing tiles, and type of backing material should be determined by qualified and experienced architects or engineers, who are well able to carry out this work and will be responsible for identifying and solving any problems. Codes of practice are determined by government bodies to ensure an acceptable standard of construction in the building industry. Copies of current codes are available through professional organizations and government departments. Failure to follow these codes of practice will increase the liability of client, architect, designer, ceramist or building contractor should any faults be revealed and traced to substandard materials or construction.

Renderings should have a thermal expansion similar to that of the ceramic cladding, so that the ceramic is not forced to crack or detach itself from the render. Stresses built up in tiles fixed to inappropriate backings may give rise to crazing even on products with correctly related glaze and body. Similarly, the rendering should not allow water to penetrate the back of the tile; absorbed water makes the tile body expand – and even more so when it freezes – and this expansion may be another cause of crazing.

Mortar should not contain more than the essential quantity of lime. If possible, lime-free mortars should be used, especially if unglazed ceramic is to be installed. Any free lime will dissolve in the water used to make up the mortar.

The water may be transmitted through the ceramic and evaporate on the surface, leaving a white scum which is almost impossible to eradicate. Washing only drives the lime back into the structure, to return as the surface dries.

Grouting and spacing. All ceramic pieces should be installed with a layer of softer material (mastic, adhesive or cement mortar) so that the expansion and contraction of the separate tiles or blocks may be accommodated by contraction or expansion of the bonding layer. Without such a barrier the forces of expansion created by changes in temperature acting from one piece to another may cause one or both to crack. Wall or floor tiles may buckle, which will lead to the tiles becoming loose upon contraction. Thin industrial tiles have spacer lugs on two or more edges, designed to crush if the tiles expand; they also facilitate setting with regular spaces between the tiles and avoid the patterning which can occur on long runs of carelessly spaced tiles. The standard technique for regulating spaces between tiles is to use material of regular thickness (pieces of thermoplastic tiles are popular) to separate each tile from its neighbour. These are removed when the tiles are securely set, and the surface then grouted. Grouting can be coloured to match the ceramic, or to contrast with it. Mortar may be similarly treated.

Metal ties should all be proof against corrosion either by treatment (galvanizing) or by the nature of the metal. Apart from weakening the metal, corrosion may stain some types of ceramic.

On-site storage should be avoided if possible. If it is unavoidable, the ceramic pieces should be stored in dry, frost-proof conditions and safe from inexperienced construction workers.

Ceramic sculpture should be designed to withstand the conditions in which it is to be placed. It should not 'frost' or otherwise deteriorate when subject to normal wear and tear. It should be fastened securely so that wind, or accidental contact, will not dislodge it. Any projecting parts should be free from faults which might cause them to break off.

If in doubt, it may be advisable to lay up glass fibre and resin, or concrete complete with galvanized wire, or mesh reinforcement on the inside, so that, in the event of the ceramic fracturing, the piece will not disintegrate.

The specification for the foundations and the method of attaching the sculpture to the ground should be decided by an engineer or architect, with due regard to the weight of the sculpture and the forces that are likely to be generated upon it by the wind.

Wall panels and tiles can be fastened by providing bolt-holes in the panel, which can be plugged with matching ceramic pieces after installation is completed. The panel may be designed so that it is assembled into the wall, particularly if this is a brick construction. In effect, such pieces are both structural and decorative.

126 Installing large faience slabs. Horizontal rods tied to the wall carry the slabs until the mortar has set, and continue to hold them in place by virtue of the key.

Small tiles may be fixed with adhesives, either thick-bed or thin-bed type. It is essential that the backing is secure if it is rendered, to avoid the possibility of it detaching and the tiles falling off. The key on the back of the tiles is often much reduced if it is designed for adhesive fixing, as the tiles are held in place by the grip of the adhesive rather than the physical shape of the hard bond. Heavier tiles, such as twin tiles, are either assembled into the wall panel at the point of manufacture, in the case of pre-cast concrete units, or assembled from the bottom layer upward, using a temporary batten to hold the tiles in place until the mortar or adhesive has set hard.

Adhesive installations are rapid and relatively unskilled, but mortar setting is skilled and expensive. If tiles are to be cut to shape round projections in the surface, or at the end of the row, the appropriate cutter should be used. Several commercial cutters are available for thin adhesive-set tiles. For thicker tiles, a circular saw with cutting disc, and water spray to trap the dust produced by the cutting, should be used. Otherwise a hammer and chisel or bolster will be necessary. The finished edge will probably be very crude, but better results can be achieved by scoring the top surface

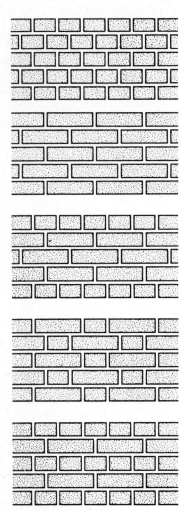

127 Type of brick bonds. *Top to bottom:* header bond, stretcher bond, facing bond, Flemish bond, English bond.

with a tungsten-tipped tool, and nibbling to this line with pliers or chisel.

Faience and terracotta blocks are structural pieces, and are assembled as stonework. With sculptural pieces that have been fired in sections, provision may be made for pegging with metal dowels designed to fit into a socket in the adjacent piece. Traditionally, these dowels were of bronze, held in with lead poured into the oversize hole when the block had been fired. Modern adhesives and resins replace the lead, and stainless or galvanized steel pegs are used instead of the bronze dowel. Each faience block is surrounded by mortar, as brickwork. Hollow blocks may be supported by an iron frame structure, the blocks being filled with straw so that the mortar or cement does not flow into the back of the block and increase the weight unnecessarily.

Floor tiles and paviors are normally set on a flat screed of cement and grouted. Thick paviors may be set on fine rubble which has been rolled to consolidate it and produce a flat, smooth surface.

Bricks may be bonded in many different patterns but the setting should follow current codes of practice.

Special or difficult shapes can be cut with a saw such as that described for cutting thick tiles. Soft-fired bricks (called rubbers) are used for facings where either the thickness of the mortar has to be reduced or special shapes (profile or on the face) created in the building. Current practice would be to produce special moulds for such bricks. The use of rubbers allows for the fine laying of bricks; bricklaying used to include these skills, which would now be very expensive to employ. The rubbing was carried out against an abrasive stone or a harder-fired brick. Water was used as a lubricant and to improve the finish of the rubbed surface. By this method irregularities in the brick face were eradicated and each brick could lie against its neighbour with only a sliver of mortar between them. This improves the weathering characteristics of brick surfaces, as thick mortar tends to wash away until repointing becomes necessary.

Appendix:
Safety in the workshop

All machinery is potentially dangerous. Never operate a machine unless you have had proper instruction in the use of it. Attend to the maintenance of machinery regularly, and keep a stock of essential spare parts so that any breakages cause no more delay than is absolutely necessary. When carrying out maintenance and repairs, ensure that machinery or kiln is switched off at the mains connection. Never open a kiln door when the kiln is on.

Working surfaces which are used for modelling clay should be covered by plastic laminate so that they may be cleaned of all clay dust at least once a day. Clean up all clay dust from floors, drying areas and working surfaces by means of a vacuum cleaner with total filtration: this is the only type of cleaner that will pass no fine dust (the most harmful if inhaled) with the exhaust air.

Always fettle clay in a fettling booth or, if this is impossible, use a room with good extraction and wear a dust filter mask.

When using resins, wear rubber gloves, follow the manufacturer's instructions and wipe up any spilled resin or hardener. Always wear a fume mask and goggles.

Keep a well stocked first-aid box and protect even the smallest scratch with a waterproof plaster.

When weighing out glaze materials use scoops or ladles and keep the containers closed at all other times, to avoid contamination. Always wash your hands after handling glaze materials, and scrub your fingernails.

Store dry materials in airtight containers outside the normal working areas. Label them carefully, as they are all of similar appearance in this state. Keep colouring materials away from other dry materials: contamination is very easy and may pass unnoticed until it is too late.

Keep plastic clay in a damp store and ensure that the various clays do not become confused with each other.

Do not drink, smoke or eat while you are working, and never use workshop containers for food or drink.

Have regular health checks, with particular reference to the presence of lead, cadmium and other heavy metals, and respiratory disease (silicosis).

Keep the telephone numbers of the emergency services, such as the doctor, hospital, ambulance and fire services, near the telephone.

Further reading

BARNARD, Julian: *Victorian Ceramic Tiles*. London, 1972
BERENDSON, Anne, *et al.*: *Tiles: a General History*. London, 1962; New York, 1967
BRITISH TERRA COTTA SOCIETY: *Terra Cotta of the Italian Renaissance*. London, 1930
DOBSON, E.: *A Rudimentary Treatise on the Manufacture of Bricks and Tiles*. London, 1850; reprinted 1971 by *Journal of Ceramic History*, No. 5
FURNIVAL, W. S.: *Leadless Decorative Tiles, Faience and Mosaic*. London, 1904
GAUNT, W. and M. Clayton Stamm: *William De Morgan*. London, 1971
VICARY, Richard: *Manual of Lithography*. London and New York, 1976
—: *Manual of Advanced Lithography*. London and New York, 1977
WOODFORDE, J.: *Bricks to Build a House With*. London and New York, 1975

Glossary

A basic compound which acts as a flux in glazes. ALKALI

The metal thread or screw which drives the clay forward in a pugmill. AUGER

A refractory material such as alumina or china clay applied to a kiln shelf to prevent glaze sticking to the shelf. BAT WASH

Blistering effect of overfiring a clay body. BLOATING

In brick or roof tile-making, a kiln in which the atmosphere is reducing. BLUE KILN

Container for mixing clay or glaze with water by means of a slow-moving paddle. BLUNGER

Mixture of clays and minerals to produce a workable material for ceramic manufacture. BODY

A wedge-shaped chisel for splitting bricks. BOLSTER

The overlapping arrangement of bricks upon one another. BOND

The degree of speckling induced in bricks and tiles by reduction firing; can be 'light' or 'heavy'. BRINDLE

A disc compounded of ceramic materials which shrinks at a predetermined rate upon firing. BULLER'S RING

Changing the physical properties of a material by heating it to red heat and allowing it to cool before use. CALCINING

The spaces between particles of clay which are filled by water when the clay is damp and plastic. CAPILLARIES

Various types of clay and ceramic material in the form of artificial fibre with a high degree of thermal insulation. CERAMIC FIBRE

A process of transferring a photographic image on to a tile by photosensitizing the glaze surface. CERMIFAX

Laying bricks so that spaces are left between them, particularly in kiln-building. CHEQUERING

CLAMP	A simple kiln made of bricks and used on temporary sites of brick-making.
CLAY MEMORY	The property which some clay forms demonstrate, in which they tend to distort towards the original form into which the clay was built or pressed before taking their final form.
CLOT	A lump of clay before it is shaped, particularly in brick-making.
COADE STONE	A type of terracotta made at the Coade Works in Lambeth between 1760 and 1830. Also known as Coade's Artificial Stone and regarded as the highest quality of terracotta ever made, in terms of durability.
CONTINUOUS KILN	A kiln which may be kept firing without the necessity of cooling the fabric in order to unpack the fired ware.
COTTLE	The wall which surrounds a model to contain the liquid plaster during mould-making. Usually a plastic material, which itself may be called 'cottling'.
CUERDA SECA	A process of painting a design in a material – usually manganese dioxide – which will separate the glazes painted in the various shapes of the design so that they do not run and mingle when fired.
DE-AIRING	The act of removing air from clay, usually by passing it through a vacuum chamber forming part of a pugmill.
DEFLOCCULANT	A compound which reduces the tendency of clay particles to attract each other and form a solid mass.
DEVITRIFICATION	The formation of crystals during the cooling of certain liquids.
DIE	A hole, of metal or other hard material, through which clay is pressed, so that it takes on the form of the hole in cross-section. In tile-making, the two faces which act upon the clay to produce a tile during pressing.
DUNTING	The cracks formed during cooling of ceramic forms when the process takes place faster than the ceramic can accommodate.
EARTHENWARE	A type of porous ceramic.
ENCAUSTIC	Inlaid pattern.
ENGOBE	A slip suitable for application over a clay form.
FAIENCE	Glazed earthenware or, in architectural ceramics, glazed terracotta.

An element or compound which acts upon others, causing them to melt when heated.	FLUX
Silica which does not act as part of the clay particles and is free to change its crystalline form when heated and cooled.	FREE SILICA
A compound of several materials which renders one or more of them easier or safer to use.	FRIT
A glassy covering applied to a ceramic material.	GLAZE
The ability of glazes to expand and contract at the same rate as the clay during heating and cooling.	GLAZE FIT
Clay which has not been fired, usually firm but not yet dry.	GREEN CLAY
Clay which has been fired and ground so that it may be used as an ingredient in a clay body to reduce the shrinkage and increase the size of the capillaries.	GROG
Placing or stacking bricks so that they may dry evenly.	HACK
A tool for cutting clay, consisting of a U-shaped metal frame with a taut cutting wire strung across the open end.	HARP
The amount of heat applied to a material in proportion to the time during which the heat is applied.	HEAT WORK
Either a shape upon which a preformed clay object may be supported while it is being modified, or the wooden frame which holds a metal template as it is moved through semi-liquid plaster during model-making.	HORSE
Machine for forming clay shapes in a spinning mould by means of a template.	JOLLY
A rubber or flexible metal scraping and smoothing tool, shaped like a kidney bean.	KIDNEY
The layering of material either in flat planes (in tile-making) around a central core (in extrusions).	LAMINATION
Earthenware covered with white (tin) glaze, usually decorated with in-glaze or on-glaze pigments.	MAIOLICA
The process of transferring a pattern from a piece of paper on to a tile or pot by piercing the lines of the pattern and then scattering finely powdered charcoal through the pinholes in the paper so that it falls on to the tile in the shape of the design outline. (Silk screens have replaced paper in modern use.) Powdered lampblack is an alternative.	POUNCING
A kiln fired with an oxidizing atmosphere in brick and tile production.	RED KILN

REDUCTION	The technique of producing an oxygen-depleted atmosphere within the firing chamber of a kiln in order to modify the metal oxides present in or upon the clay so that they may produce different colours and textures from those that would result with an oxygen-rich atmosphere.
REFRACTORY	Resistant to the melting effects of heat.
SAGGAR	A refractory box used to protect clay forms from direct contact with the kiln fire.
SCINTLING	Arranging tiles or bricks in herring-bone patterns so that when the next layer is set in the opposite direction air spaces are left and the clay may continue to dry or, if in a kiln, the reduction atmosphere may reach every article equally.
SINTERING	Initial surface bonding of clay particles during firing.
SLEDGE	An accurately engineered tool which will hold a template and run true along an edge to produce strips of plaster in the form of the template.
SLIP	Mixture of clay and water.
SLIP TRAILER	A reservoir and fine nozzle for decorating clay.
SPALLING	Cracking upon cooling, due to either thermal shock or differential thermal expansion.
SPIT-OUT	The shelling of the surface of a ceramic form caused by pressure generated within it either by the rehydration of calcium ('plaster spit-out') or iron pyrites or by moisture trapped in glazed earthenware forcing its way through a transfer applied over the glaze.
STONEWARE	Impervious ceramic with much of the material undissolved but bound by a glassy matrix.
SURFACE DRYING	Drying of the surface of clay in contact with plaster more rapidly than the clay beneath the surface.
TERRACOTTA	Unglazed porous ceramic, sometimes with a coarse texture and often made of red-burning clay.
TESSELLATION	A pattern of one or more shapes repeated over a surface so that they fit perfectly without spaces between the shapes.
TRAVELLING TABLE	A surface which moves at the same speed at which clay is extruded from a pugmill, so that cutting wires may pass through the clay to cut it vertically.

178

A continuous kiln through which the ware to be fired is passed on kiln cars passing through heating, maturing and cooling zones. TUNNEL KILN

The forming of clay or plaster by spinning it upon either a vertical or a horizontal lathe and shaping it with templates or tools. TURNING

The process of heating a material until it melts into a glass-like substance. In ceramics this process is arrested before the material loses its shape. VITRIFICATION

Mixing or kneading of clay to produce a homogeneous mass. WEDGING

Index